Proceedings of the Third International Colloquium on Differential Equations

Proceedings of the
Third International Colloquium on Differential Equations

Plovdiv, Bulgaria, 18 - 22 August 1992

Editors: **D. Bainov and V. Covachev**

///VSP///

Utrecht, The Netherlands, 1993

VSP BV
P.O. Box 346
3700 AH Zeist
The Netherlands

First published in 1993

ISBN 90-6764-153-7

Printed in Great Britain by Bookcraft (Bath) Ltd., Midsomer Norton.

CONTENTS

Preface

The Third International Colloquium on Differential Equations was organized by UNESCO and Plovdiv Technical University with the help of the Austrian Mathematical Union, the Canadian Mathematical Society, Hamburg University, Institute of Mathematics of the Bulgarian Academy of Science, Kyushu University, the London Mathematical Society, Plovdiv University 'Paissii Hilendarski', Technical University-Berlin, Union of the Bulgarian Mathematicians-Plovdiv Section, and sponsored by Bulgarian Mails and Telecommunications Ltd., and the firms 'Eureka' and 'Microtechnika Ltd.' It was held on August 18-22, 1992 in Plovdiv, Bulgaria. Over 200 participants from 43 countries took part in its work. Round tables were organized on 'Impulse Differential Equations', Non-linear Differential Equations' and 'Differential Equations with Maxima', resulting in fruitful discussions among their participants. The subsequent colloquia will take place each year from 18-22 August in Plovdiv, Bulgaria.

The address of the organising committee is:
Stoyan Zlatev, Mathematical Faculty of the Plovdiv University, Tsar Assen Str. 24, Plovdiv 4000, Bulgaria.

The Editors

Third International Colloquium on Differential Equations pp. 1-12 (1993)
D. Bainov and V. Covachev (Eds)

TIME DELAY MODELS OF POPULATION GROWTH WITH STAGE STRUCTURE

WALTER AIELLO
Department of Mathematics, University of Alberta
Edmonton, Alberta, Canada T6G 2G1

ABSTRACT

Models of population growth incorporating immature and mature stages of growth are investigated. In particular, a comparison is made between single species models of populations where the time to maturity is a fixed constant, and where the time to maturity is a function of population concentration.

INTRODUCTION

Modelling single species growth has a long history in the area of mathematical ecology, where modern developments can be said to have started with the work of Malthus in 1798. The logistic equation,

$$\dot{x}(t) = rx(t)\left(\frac{1 - x(t)}{k}\right) \quad t \geq 0$$

plays an important role, where $x(t)$ is the concentration of individuals in the population, r is an instrinsic growth rate, and k is the environmental carrying capacity. Solutions of the logistic equation are monotone functions of time t, increasing or decreasing to the limit $x(t) = k$ if $x(0) > 0$. The literature on this subject is quite extensive, for example see Freedman [1].

The existence of features such as gestation in natural systems lead investigators to introduce time delays into the logistic equation. One form of the logistic equation with time delays,

$$\dot{x}(t) = rx(t)\left(\frac{1 - x(t-\tau)}{k}\right) \quad t \geq 0$$

is known as Hutchinson's equation. Here $r, k,$ and τ are positive constants. Hutchinson's equation has been investigated by May [2], Kakutani and Markus [3] and by Zhang and Gopalsamy [4].

In our investigations, we consider a population that is divided into immature and mature stages of growth. The equation for this model thus appears in two parts, one for the immature stage and one for the mature stage of growth.

CONSTANT TIME DELAY MODEL - DEVELOPMENT

When the time to maturity remains constant, the model takes on the form,

$$\begin{aligned} \dot{x}_i(t) &= \alpha x_m(t) - \gamma x_i(t) - \alpha e^{-\gamma\tau} x_m(t-\tau), \quad t > 0 \\ \dot{x}_m(t) &= \alpha e^{-\gamma\tau} x_m(t-\tau) - \beta x_m^2(t). \end{aligned} \tag{1}$$

Here α is the intrinsic birth rate of the population, γ is the intrinsic death rate of immatures, and β is the logistic death rate of matures. For $0 \leq t \leq \tau$, $x_m(t-\tau)$ takes on the value of an initial function $\phi(t)$ that is defined on the interval $[-\tau, 0]$. To assure continuity of initial conditions we assume that,

$$x_i(0) = \int_{-\tau}^{0} \alpha\phi(t) e^{\gamma\tau} dt.$$

Thus we see that introduction of new members into the population is by birth of new immatures, and that those immatures born at time $t-\tau$ who survive to time t leave the immature population and pass into the mature population.

Assuming the initial function to be continuous (for mathematical reasons) and positive (for biological reasons) we prove that for system (1) the following are true:

1) Solutions $(x_i(t), x_m(t))$ exist and are uniquely continuable,
2) Solutions $(x_i(t), x_m(t))$ are positive for all t,
3) Solutions $(x_i(t), x_m(t))$ are bounded.

The proof of 1) follows from the theory of existence and uniqueness of delay differential systems as outlined in Driver [5]. Properties 2) and 3) are proven in Aiello and Freedman [6].

EQUILIBRIA AND STABILITY FOR CONSTANT DELAY CASE

There are two possible equilibria, one at the origin, $E_0(0;0)$, and in the interior, $\widehat{E}(\widehat{x}_i, \widehat{x}_m)$. In Aiello and Freedman [6] it is shown that $\widehat{E}$ exists, and that $\widehat{x}_i$ and

$\widehat{x}_m$ satisfy,

$$\widehat{x}_i = \alpha^2\beta^{-1}\gamma^{-1}e^{-\gamma\tau}(1 - e^{-\gamma\tau})$$
$$\widehat{x}_m = \alpha\beta^{-1}e^{-\gamma\tau}.$$

The variational system about E_0 gives us a characteristic equation $(\lambda + \gamma)(\lambda - \alpha e^{-\tau(\lambda+\gamma)}) = 0$. This equation has at least one real, positive eigenvalue and at least one real, negative eigenvalue. Thus E_0 is a "saddle" point.

Concerning the interior equilibrium $\widehat{E}(\widehat{x}_i, \widehat{x}_m)$, in Aiello and Freedman [6], it is proven that this equilibrium is globally asymptotically stable.

SUMMARY OF CONSTANT DELAY CASE

We have shown that the model for a population with two growth stages and fixed time delay has a globally asymptotically stable interior equilibrium. The results of a computer simulation of several solutions with different initial conditions is illustrated in figure number 1.

STATE DEPENDENT TIME DELAY MODEL - DEVELOPMENT

Observation of the population dynamics of certain mammalian species such as seals and whales in the Antarctic Ocean appears to indicate that the time to maturity is shortened with decreased populations and the resulting increase in food supply that becomes available to each member of the species. Hence we consider a model of single species growth where the time to maturity is a monotone nondecreasing function of members of the species present.

Thus for our model we apply the system,

$$\begin{aligned}\dot{x}_i(t) &= \alpha x_m(t) - \gamma x_i(t) - \alpha e^{-\gamma\tau(z)}x_m(t - \tau(z)) \\ \dot{x}_m(t) &= \alpha e^{-\gamma\tau(z)}x_m(t - \tau(z)) - \beta x_m^2(t)\end{aligned} \tag{2}$$

where $x_m(t) = \varphi_m(t) \geq 0$ for $-\tau_m \leq t \leq 0$, and where $\alpha > 0,\ \beta > 0,\ \gamma > 0,\ \tau_m \leq \tau(z) \leq \tau_M$. Here $z(t) = x_i(t) + x_m(t)$.

The variable delay system is similar to the previous constant delay model except that the time delay is now a function of $z(t)$ rather than a constant. The time delay

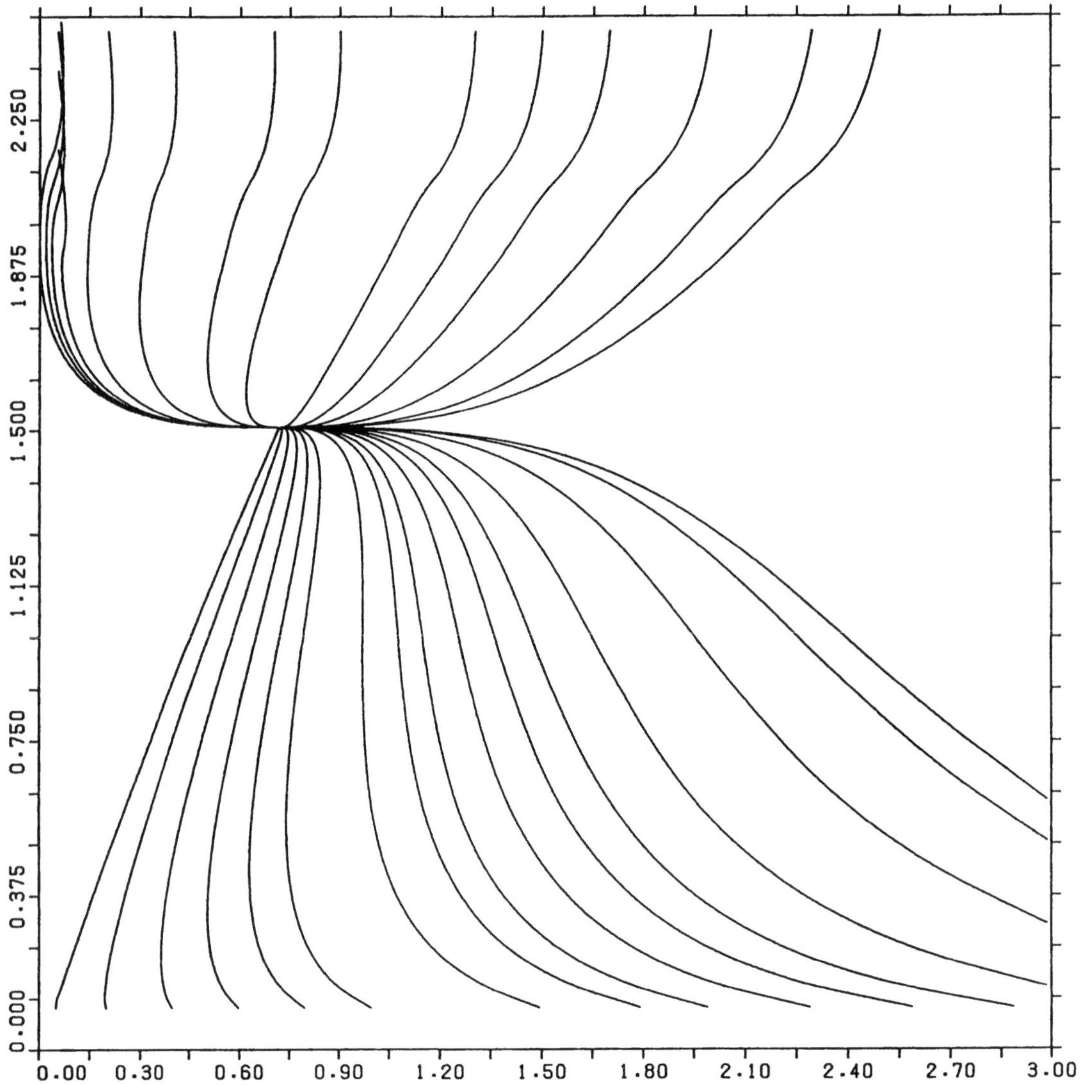

Figure 1. Globally Asymptotically Stable Equilibrium

function $\tau(z)$ is assumed to be nondecreasing, and it is further assumed to satisfy,

$$\lim_{z\to 0+} \tau(z) = \tau_m > 0$$
$$\lim_{z\to\infty} \tau(z) = \tau_M < \infty.$$

We further assume that $d\tau/dz = \tau'(z)$ exists, so that $\tau'(z) \geq 0,\ 0 \leq \tau_m \leq \tau(z) \leq \tau_M < \infty$.

For initial conditions, we have a given mature population $\varphi_m(t)$ defined on $-\tau_M \leq t \leq 0$. From this we are able to derive an immature population at time $t = 0$. To properly define $x_i(0)$ we need to make certain assumptions.

First, since $x_i(t)$ represents the accumulated survivors of those immatures who were born between times $t - \tau(z(t))$ and t, the expression $(t - \tau(z(t)))$ must be strictly increasing in order to avoid the problem of matures reverting back into immatures, or of immatures being resurrected from the dead. Thus we must further assume that the function $t - \tau(z(t))$ be a strictly increasing function of t. This will be true if the following hold:

1) $x_i(t) > 0$ for all t,
2) $0 \leq \tau'(z) < 4\beta/\alpha^2$, where $\tau'(z) = d\tau(z)/dz$.

We must assume 2) to be true, and we prove in the sequel that 1) is true given certain assumptions on initial conditions.

Now, given these assumptions on $\tau(z(t))$, and given that $\varphi_m(t) > 0$ on $-\tau_m \leq t \leq 0$, we now go on to derive $x_i(0)$.

Let us make the following definition:

$$\tau_s = \inf\{\tau_u : \tau_u = \tau\big(x_m(0) + \int_{-\tau_u}^{0} \alpha x_m(s) e^{\gamma s} ds\big)\}.$$

Here $x_m(0) + \int_{-\tau_u}^{0} \alpha x_m(s) e^{\gamma s} ds$ represents the number of matures present at time $t = 0$ plus those immatures who were born between times $-\tau_u$ and 0 and who survived to time $t = 0$. We then define $x_i(0) = \int_0^{\tau_s} \alpha \varphi_m(s - \tau_s) e^{\gamma(s-\tau_s)} ds$. Thus $x_i(0)$ represents the accumulated survivors of those members of the immature population born between time $-\tau_s$ and 0. Note that for values of t, $-\tau_s \leq t \leq 0$ we understand that $x_m(t) = \varphi_m(t)$ and that $x_i(0) = \varphi_i(0)$. Note also that $\tau(z(0)) = \tau_s$.

In this manner, continuity of initial conditions is assured.

We are now able to show that the model makes good biological sense by proving the following:

THEOREM. *Let* $\varphi_m(t) > 0$ *for* $-\tau_M \le t \le 0$. *Then* $x_m(t) > 0$ *for all* $t > 0$.

THEOREM. *Let* $\varphi_m(t) > 0$ *for* $-\tau_m \le t \le 0$. *Then there exists a* $\delta_m = \delta_m(\varphi_m) > 0$ *such that* $x_m(t) \ge \delta_m$ *for all* $t \ge 0$.

THEOREM. *Let* $\varphi_m(t) > 0$ *for* $-\tau_m \le t \le 0$. *Then there exists a* $\Delta_m = \Delta_m(\varphi_m) > 0$ *such that* $x_m(t) \le \Delta_m$ *for all* $t > 0$.

Thus if the initial function $\varphi_m(t) > 0$, then $0 < \delta_m(\varphi_m) \le x_m(t) \le \Delta_m(\varphi_m) < \infty$ holds.

We also have for $x_i(t)$,

THEOREM. *Let* $\varphi_m(t) > 0$ *for* $-\tau_m \le t \le 0$. *Then there exists a* $\Delta_i(\varphi_m) = x_i(0) + \alpha\gamma^{-1}\Delta_m$ *such that* $x_i(t) \le \Delta_i$ *for* $t \ge 0$.

For proofs of these theorems, we refer to Aiello and Freedman [7].

Now all we need for the model to make good biological sense is for $x_i(t)$ to remain positive when $\varphi_m(t) > 0$ on $-\tau_s \le t \le 0$. To prove this, we make some additional assumptions on either the initial function $\varphi_m(t)$ or on the delay function $\tau(z)$.

THEOREM. *Suppose* $\tau'(z) < 4\beta/\alpha^2$ *is satisfied and that* $\tau'(z) \ge 0$ *is small enough so that the inequality*

$$\delta_m \int_{t-\tau_m}^{t} e^{\gamma s} ds > \Delta_m \int_{-\tau_s}^{t-\tau_m} \frac{\alpha\tau'(z)}{4\beta - \alpha^2\tau'(z)} ds$$

holds for all values of t. *Then* $x_i(t) > 0$ *for all* $t > 0$.

In addition we have,

THEOREM. *Suppose* $e^{-\gamma\tau_m} \le \delta_m/\Delta_m$. *Then* $x_i(t) > 0$ *for all* $t > 0$.

The first of these two existence theorems seems to indicate that as $\tau'(z)$ increases the set of admissible initial functions becomes more restricted, becoming null as $\tau'(z)$ approaches $\tau'(z) = 4\beta/\alpha^2$, which is the point at which the possibility of resurrection arises.

The second theorem seems to indicate, however, that no matter what $\tau'(z)$ might do, there will always be some initial function that will give a solution where $x_i(t) > 0$ for all t. Note here that δ_m and Δ_m are functions of the initial function $\varphi_m(t)$.

For proofs of these theorems we again refer to Aiello and Freedman [7]. We also mention at this point that we cannot find a nonzero bound below for $x_i(t)$.

EQUILIBRIA AND STABILITY - STATE DEPENDENT CASE

The state dependent model has an equilibrium $E_0(0,0)$ and at least one interior equilibrium $\widehat{E}(\widehat{x}_i, \widehat{x}_m)$. More than one interior equilibrium may exist, but we can give sufficient conditions for uniqueness:

THEOREM. *If any one of*

i) $\gamma > \alpha$,

ii) $\widehat{x}_m < \frac{\alpha+\gamma}{2\beta}$

iii) $\tau'(\widehat{z}) < \frac{1}{\widehat{x}_m(2\beta\widehat{x}_m - \alpha - \gamma)}$

is true, then $\widehat{E}$ *is unique.*

Aiello and Freedman [7] contains a discussion on the existence of equilibria and a proof of the theorem.

Before engaging in any local stability analysis, we note that linearized stability theory for state dependent delay systems is not yet completely developed. Thus the following local stability analysis is only formal in nature.

Let $E^*(x_i^*, x_m^*)$ be an arbitrary equilibrium. The variational system about E^* lends to the characteristic equation

$$\det \begin{vmatrix} \lambda + \gamma - \xi^* & \alpha e^{-\tau(z^*)(\lambda+\gamma)} - \alpha - \xi^* \\ \xi^* & \lambda + 2\beta x_m^* + \xi^* - \alpha e^{-\tau(z^*)(\lambda+\gamma)} \end{vmatrix} = 0$$

where $\xi^* = \alpha\gamma e^{-\gamma\tau(z^*)}\tau'(z^*)x_m^*$.

At $E_0(0,0)$ $\xi^* = 0$ and the characteristic equation reduces to

$$(\lambda + \gamma)(\lambda - \alpha e^{-\tau_m(\lambda+\gamma)}) = 0.$$

This has at least one real positive and at least one real negative solution, so E_0 is an unstable "saddle" point.

From the characteristic equation about an interior equilibrium $\widehat{E}$ we can show that following to be true:

THEOREM. *Let* $\tau'(\widehat{z}) = 0$. *Then* $\widehat{E}$ *is locally asymptotically stable.*

THEOREM. *If either*

i) $\widehat{x}_m \le \frac{\alpha+\gamma}{2\beta}$ or

ii) $\widehat{x}_m > \frac{\alpha+\gamma}{2\beta}$ and $\tau'(\widehat{z}) \le \frac{3}{4\widehat{x}_m(2\beta\widehat{x}_m - \alpha - \gamma)}$

is true, then $\widehat{E}$ *is locally asymptotically stable.*

For further discussion of this topic and the proofs of the theorems, refer to Aiello and Freedman [7].

GLOBAL BEHAVIOUR OF SOLUTIONS

In this section we state some results on the global behaviour of solutions which are proved in Aiello and Freedman [7].

Our first result is that if $x_m(t)$ stays above or below certain values long enough, then $x_m(t)$ will stay above or below those values for all time.

THEOREM. *Let* $(x_i(t), x_m(t))$ *be a solution of the state dependent system. Then,*

i) If there exists $t_1 \ge -\tau_m$ *such that* $x_m(t) \le \alpha\beta^{-1}e^{-\gamma\tau_m}$ *for* $t_1 \le t \le t_1 + \tau_M$, *then* $x_m(t) \le \alpha\beta^{-1}e^{-\gamma\tau_m}$ *for all* $t \ge t_1$.

ii) If there exists $t_2 \ge -\tau_m$ *such that* $x_m(t) \ge \alpha\beta^{-1}e^{-\gamma\tau_M}$ *for* $t_2 \le t \le t_2 + \tau_M$, *then* $x_m(t) \ge \alpha\beta^{-1}e^{-\gamma\tau_M}$ *for all* $t \ge t_1$.

This immediately leads to the following corollary:

COROLLARY. *If the initial function* $\varphi_m(t) < \alpha\beta^{-1}e^{-\gamma\tau_m}$, $-\tau_M \le t \le 0$, *then* $x_m(t) \le \alpha\beta^{-1}e^{-\gamma\tau_m}$ *for all* $t \ge 0$, *and if* $\varphi_m(t) > \alpha\beta^{-1}e^{-\gamma\tau_M}$, $-\tau_M \le t \le 0$, *then* $x_m(t) \ge \alpha\beta^{-1}e^{-\gamma\tau_M}$ *for all* $t \ge 0$.

Finally, we state some asymptotic results.

THEOREM. *Let* $(x_i(t), x_m(t))$ *be a solution of the state dependent system. Then*

$$\alpha\beta^{-1}e^{-\gamma\tau_M} \le \varliminf_{t\to\infty} x_m(t) \le \varlimsup_{t\to\infty} x_m(t) \le \alpha\beta^{-1}e^{-\gamma\tau_m}.$$

THEOREM. *Let* $(x_i(t), x_m(t))$ *solve the state dependent system. Then* $\overline{\lim_{t\to\infty}}\, x_i(t) \leq \alpha^2\beta^{-1}\gamma^{-1}(e^{-\gamma\tau_m} - e^{-2\gamma\tau_M})$.

THEOREM. *Let* $\tau_M < 2\tau_m$. *Let* $(x_i(t), x_m(t))$ *be a solution of the state dependent system. Then* $\underline{\lim}_{t\to\infty}\, x_i(t) \geq \alpha^2\beta^{-1}\gamma^{-1}(e^{-\gamma\tau_M} - e^{-2\gamma\tau_m})$.

Note that the last theorem does not give an explicit lower bound for $x_i(t)$ as $t \to \infty$ if the difference between τ_M and τ_m is larger than τ_m.

CONCLUSIONS

We see that for a state dependent time delay model the global asymptotic stability of the constant time delay model is replaced by a region of attraction in the $x_i - x_m$ plane. As the difference between τ_m and τ_M decreases, this region shrinks, and when $\tau_m = \tau_M$ the region becomes a point, which is just the globally asymptotically stable equilibrium of the constant case.

As well, the state dependent case admits the possibility of multiple equilibria. Thus the single globally asymptotically stable equilibrium point of the conventional logistic model and the constant time delay case becomes replaced with the possibility of multiple equilibria. Figure 2 shows the results of a computer simulation of the state dependent time delay model with various initial conditions. Here we apparently have one asymptotically stable equilibrium. We were unable to provide an example of multiple equilibria without violating the condition that $\tau'(z) < 4\beta/\alpha^2$. Thus the example for which we offer a computer simulation would display resurrection, rendering it useless as a biological model. However, for the sake of the mathematics involved we offer the simulation in figure 3.

ACKNOWLEDGEMENT

The author wishes to thank Dr. H. Freedman of the Department of Mathematics for his most helpful comments, suggestions, and discussions.

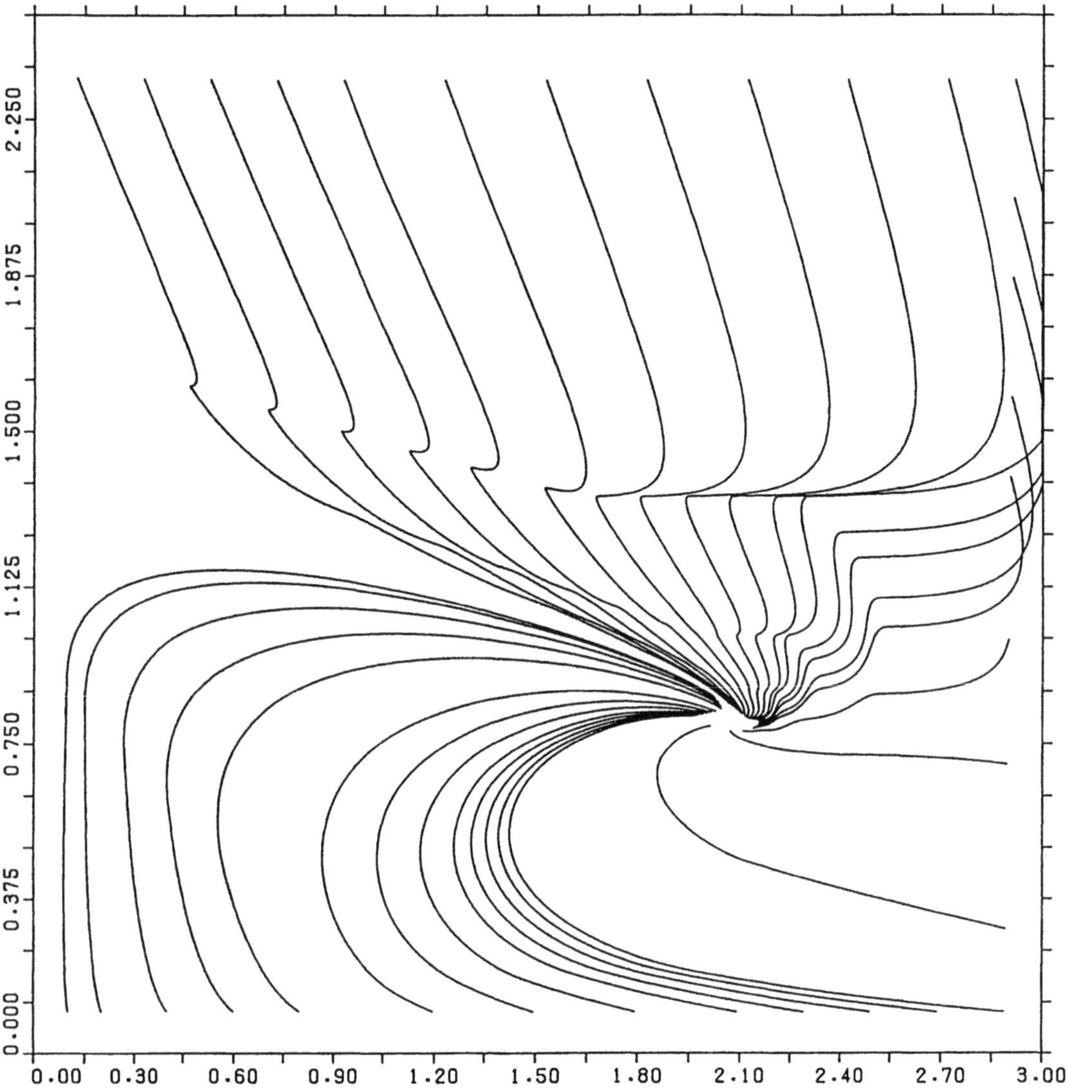

Figure 2. Single Equilibrium
State Dependent Time Delay Model

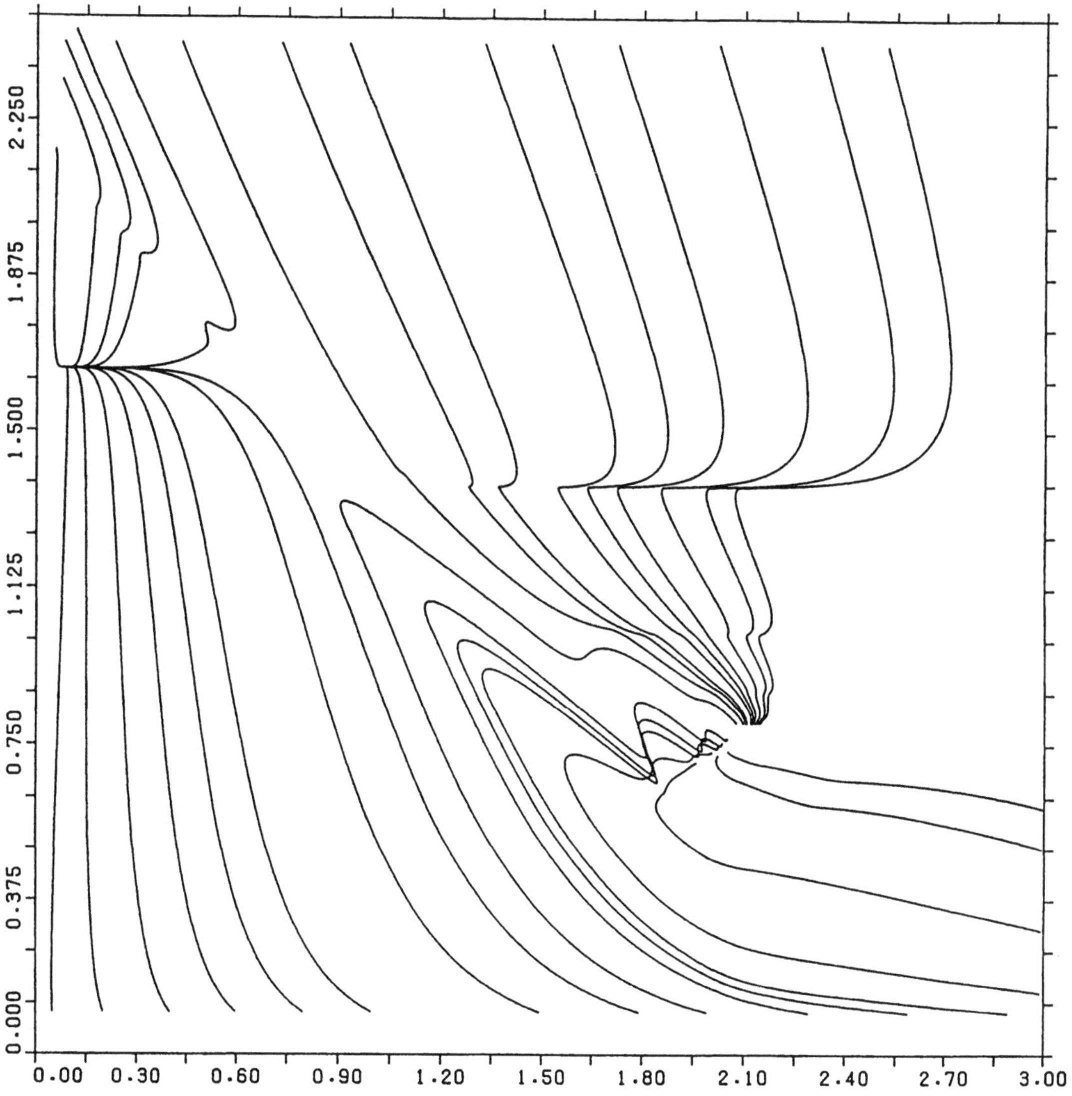

Figure 3. Multiple Equilibria
State Dependent Time Delay Model

REFERENCES

1. H.I. Freedman, Deterministic Mathematical Models in Population Biology, HIFR Consulting Ltd., (1987), pp. 8-11.

2. R.M. May, Ecology, 54, pp. 315-325, (1973).

3. S. Kakutani and L. Markus, in "Contributions to the Theory of Nonlinear Oscillations, V. 4," Annals of Mathematics Study, 41, Princeton University Press (1958).

4. B. Zhang and K. Gopalsamy, Quart. Appl. Math., 46, pp. 267-273, (1988).

5. R.D. Driver, Ordinary and Delay Differential Equations, Springer, New York, (1977)

6. W.G. Aiello and H.I. Freedman, Math. Biosciences, 101, pp. 139-153, (1990).

7. W.G. Aiello and H.I. Freedman, SIAM J. Applied Math., V 52, No. 3, pp. 855-869, (1992).

Third International Colloquium on Differential Equations pp. 13-22 (1993)
D. Bainov and V. Covachev (Eds)

Sufficient Conditions for Point Dissipative Quadratic Nonlinear Dynamical Systems

Anil K. Bose, Alan S. Cover and James A. Reneke
Department of Mathematical Sciences, Clemson University, Clemson, SC 29634-1907

Abstract
Conditions are investigated for systems of the form $M\mathbf{x}' = N + A\mathbf{x} + \mathbf{f}(\mathbf{x})$, where f is quadratic, which yield a point dissipative system. Application of the conditions are made to the problem of existence of linear feedback controls $\mathbf{u} = K\mathbf{x}$ for systems of the form $M\mathbf{x}' = N + A\mathbf{x} + \mathbf{f}(\mathbf{x}) + B\mathbf{u}$ which force the system to be point dissipative. The basic results have extensions to more general classes of systems.

1. Introduction

We are concerned with a class of nonlinear n-dimensional systems of the form $M\mathbf{x}' = N + A\mathbf{x} + \mathbf{f}(\mathbf{x})$, where M is a positive definite matrix and the nonlinear term $\mathbf{f}(\mathbf{x})$ is quadratic of the form

$$\mathbf{f}(\mathbf{x}) = \begin{bmatrix} \mathbf{x}^T C_1 \mathbf{x} \\ \\ \\ \mathbf{x}^T C_n \mathbf{x} \end{bmatrix}$$

The $n \times n$ matrices $\{C_i\}$ are symmetric with the orthogonality property $\mathbf{x}^T\mathbf{f}(\mathbf{x}) = 0$ for all $\mathbf{x}$. Functions $\mathbf{f}$ with the orthogonality property are said to be conservative. We start with an investigation of conditions on the $n \times n$ matrix A and $\mathbf{f}$ sufficient to have the system point dissipative, i.e., which guarantee the existence of a bounded region in R^n which every trajectory of the system eventually enters and remains within. In other words, conditions which imply the system has a bounded attractor.

As an example of a point dissipative system of the form given above, we cite the well known Lorenz system [5] $\mathbf{x}' = A\mathbf{x} + \mathbf{f}(\mathbf{x})$, where

$$A = \begin{bmatrix} -a & a & 0 \\ \gamma & -1 & 0 \\ 0 & 0 & -b \end{bmatrix}, \quad a > 0,\ \gamma > 0,\ b > 0$$

$$\mathbf{f}(\mathbf{x}) = \begin{bmatrix} 0 \\ -x_1 x_3 \\ x_1 x_2 \end{bmatrix}, \quad \mathbf{x} = \begin{bmatrix} x_1 \\ x_2 \\ x_3 \end{bmatrix}$$

Note that $\mathbf{f}(\mathbf{x})$ has the orthogonality property. Furthermore, if $\mathbf{u}$ is a nontrivial member of the zero set of $\mathbf{f}(\mathbf{x})$, i.e., $\mathbf{u} = (u_1, 0, 0)^T$ or $\mathbf{u} = (0, u_2, u_3)^T$ then $\mathbf{u}^T A \mathbf{u} < 0$. Both of these properties are preserved under change of coordinate transformations.

When n = 2 or 3, a sufficient condition for the system to be point dissipative is that $\mathbf{u}^T A \mathbf{u} < 0$ for nontrivial $\mathbf{u}$ in the zero set of $\mathbf{f}(\mathbf{x})$. The conjecture that this condition is sufficient for n-dimensional systems is unresolved except for special cases, i.e., systems where the zero set of $\mathbf{f}(\mathbf{x})$ satisfies additional conditions.. Theorem 1 extends the n-dimensional result in [2].

Since linear feedback does not disturb the zero set of $\mathbf{f}(\mathbf{x})$, our feeling is that linear control minimally modifies the structure of the uncontrolled system. Our goal is to produce a linear feedback so the controlled system either has the origin as an asymptotic stable point or is point dissipative. No use is made of nonlinear feedback.

The extensions in Section 4 relax the conditions on $\mathbf{f}(\mathbf{x})$, i.e., $\mathbf{f}(\mathbf{x})$ need not be conservative or quadratic.

2. Basic results

It can be shown [2] that for quadratic $\mathbf{f}(\mathbf{x})$ with the orthogonality property the set $Z_f = \{\mathbf{x} \mid \mathbf{f}(\mathbf{x}) = 0\}$ contains at least a 1-dimensional subspace of R^n. For each vector $\alpha^T = (\alpha_1, \alpha_2, \ldots, \alpha_n)$, we define the $n \times n$ matrix $C(\alpha)$ as follows:

$$C(\alpha) = \sum_{i=1}^{n} \alpha_i C_i - \frac{A + A^T}{2}$$

Our first result is the following lemma.

<u>Lemma</u> 1. If there exists an α so that the matrix $C(\alpha)$ is positive definite, then the system $M\mathbf{x}' = N + A\mathbf{x} + \mathbf{f}(\mathbf{x})$ is point dissipative.

This lemma is proven by a standard argument using the Lyapunov function $V(x) = \frac{1}{2}(\mathbf{x} - \alpha)^T M(\mathbf{x} - \alpha)$. For large $\|\mathbf{x}\|$ the time derivative of $V(\mathbf{x}(t))$ is dominated by $-\mathbf{x}^T(t)C(\alpha)\mathbf{x}(t)$. Vectors α for which $C(\alpha)$ is positive definite are said to be admissible for the system $M\mathbf{x}' = N + A\mathbf{x} + \mathbf{f}(\mathbf{x})$.

A condition on the matrix A and $\mathbf{f}(\mathbf{x})$ which guarantees the existence of admissible α's is the topic of our next result. The condition $\mathbf{u}^T A \mathbf{u} < 0$ for all nontrivial $\mathbf{u}$ in Z_f is necessary for the existence of an admissible α. We have shown [2] the condition is sufficient when n is 2 or 3. We have also shown [3, 4] that this condition is sufficient when Z_f is an (n - 1)-dimensional subspace of R^n. The next result weakens this equality to inclusion.

<u>Theorem</u> 1. Consider the system $M\mathbf{x}' = N + A\mathbf{x} + \mathbf{f}(\mathbf{x})$, where $\mathbf{f}$ is conservative. If Z_f contains an (n - 1)-dimensional subspace of R^n then $\mathbf{u}^T A \mathbf{u} < 0$ for all nontrivial $\mathbf{u}$ in Z_f if and only if there is an admissible α for the system.

Notation

We start by considering the conservative quadratic function $\mathbf{f}(\mathbf{x})$.

$$\mathbf{f}(\mathbf{x}) = \begin{bmatrix} \mathbf{x}^T C_1 \mathbf{x} \\ \mathbf{x}^T C_2 \mathbf{x} \\ \cdot \\ \cdot \\ \cdot \\ \mathbf{x}^T C_n \mathbf{x} \end{bmatrix} \quad \text{and} \quad \mathbf{x}\, Q\, \mathbf{y} = \begin{bmatrix} \mathbf{x}^T C_1 \mathbf{y} \\ \mathbf{x}^T C_2 \mathbf{y} \\ \cdot \\ \cdot \\ \cdot \\ \mathbf{x}^T C_n \mathbf{y} \end{bmatrix}$$

Q is a bilinear function and $\mathbf{f}(\mathbf{x}) = \mathbf{x}Q\mathbf{x}$. We denote the zeros of $\mathbf{f}$ by Z_f. The vector space generated or spanned by $\mathbf{u}_1, \mathbf{u}_2, \ldots, \mathbf{u}_k$ is denoted by $S(\mathbf{u}_1, \mathbf{u}_2, \ldots, \mathbf{u}_k)$

Lemma 2. If $S(\mathbf{u}_1, \mathbf{u}_2, \ldots, \mathbf{u}_k) \subset Z_f$ and $\mathbf{u}_1, \mathbf{u}_2, \ldots, \mathbf{u}_k$ are linearly independent and $\mathbf{u}^T A \mathbf{u} <$ 0 for all non-trival $\mathbf{u}$ in Z_f, then the matrix $B = [\, - \mathbf{u}_i^T A \mathbf{u}_j\,]_{k \times k}$, is positive definite.

Lemma 3. The subspace $S(\mathbf{u}_1, \mathbf{u}_2, \ldots, \mathbf{u}_k) \subseteq Z_f$ if and only if $\mathbf{u}_i Q \mathbf{u}_j = 0$ for all $i, j = 1, 2, \ldots,$ k.

Lemma 4. Either $Z_f = S(\mathbf{u}_1, \mathbf{u}_2, \ldots, \mathbf{u}_{n-1})$ or $Z_f = S(\mathbf{u}_1, \mathbf{u}_2, \ldots, \mathbf{u}_{n-1}) \cup S(\mathbf{v}_1, \mathbf{v}_2, \ldots, \mathbf{v}_h)$ where the $\mathbf{u}$'s and $\mathbf{v}$'s are linearly independent and $h \leq n - 1$.

3. Control results

For our next result we consider nonlinear control systems of the form $M\mathbf{x}' = N + A\mathbf{x} + \mathbf{f}(\mathbf{x}) + B\mathbf{u}$, where B is an $n \times m$ matrix and $\mathbf{u}$ is an m-vector. Two types of general behavior are investigated:

1) Existence of a linear feedback control $\mathbf{u} = K\mathbf{x}$ so that the system $M\mathbf{x}' = N + (A + BK)\mathbf{x} + \mathbf{f}(\mathbf{x})$ has the origin as a global asymptotic stable point.

2) Existence of a linear feedback control $\mathbf{u} = K\mathbf{x}$ so that the system $M\mathbf{x}' = N + (A + BK)\mathbf{x} + \mathbf{f}(\mathbf{x})$ is point dissipative.

Let S(B) be the column space of B and $B^{\perp}$ be the orthogonal complement of S(B).

Lemma 5. If $\mathcal{M}$ is a nonempty subset of $\{\mathbf{x} \mid \mathbf{x} \in B^{\perp}$ and $\|\mathbf{x}\| = 1\}$ such that $\beta^T A \beta < 0$ for β in $\mathcal{M}$ and $\mathcal{N}$ is a nonempty subset of $\{\, \mathbf{x} \mid \mathbf{x} \in S(B)$ and $\|\mathbf{x}\| = 1\}$ then there is a negative number τ such that $(a\beta + bv)^T(A + \tau BB^T)(a\beta + bv) < 0$ for β in $\mathcal{M}$, v in $\mathcal{N}$ and scalars a and b, not both zero.

Theorem 2. There exists a matrix K such that the system $M\mathbf{x}' = (A + BK)\mathbf{x} + \mathbf{f}(\mathbf{x})$ has the origin as a global asymptotic stable point if $\mathbf{u}^T A\mathbf{u} < 0$ for each nontrivial vector $\mathbf{u}$ which is orthogonal to the column space of B.

Example for Theorem 2. Consider the control system $\mathbf{x}' = A\mathbf{x} + \mathbf{f}(\mathbf{x}) + B\mathbf{u}$, where

$$A = \begin{bmatrix} 3 & 1 & -4 \\ 2 & -1 & 0 \\ -4 & 0 & 1 \end{bmatrix}, \ \mathbf{f}(\mathbf{x}) = \begin{bmatrix} 0 \\ -xz \\ xy \end{bmatrix} \text{ and } B = \begin{bmatrix} 1 \\ 0 \\ -1 \end{bmatrix}$$

Here $B^{\perp} = S\left\{\begin{bmatrix} 0 \\ 1 \\ 0 \end{bmatrix}, \begin{bmatrix} 1 \\ 0 \\ 1 \end{bmatrix}\right\}$ and $Z_f = S\left\{\begin{bmatrix} 0 \\ 1 \\ 0 \end{bmatrix}, \begin{bmatrix} 0 \\ 0 \\ 1 \end{bmatrix}\right\} \cup S\left\{\begin{bmatrix} 1 \\ 0 \\ 0 \end{bmatrix}\right\}$. Note that $\mathbf{u}$ in $B^{\perp}$ and $\mathbf{u} \neq 0$ implies $\mathbf{u}^T A\mathbf{u} < 0$. Take $K = \tau B^T$. Then

$$A + BK = A + \tau BB^T = \begin{bmatrix} 3+\tau & 1 & -4-\tau \\ 2 & -1 & 0 \\ -4-\tau & 0 & 1+\tau \end{bmatrix}$$

The Hermitian part of $A + \tau BB^T$ is given by

$$\begin{bmatrix} 3+\tau & \frac{3}{2} & -4-\tau \\ \frac{3}{2} & -1 & 0 \\ -4-\tau & 0 & 1+\tau \end{bmatrix}$$

which is negative definite when $\tau = -7$. Hence the system $\mathbf{x}' = (A - 7BB^T)\mathbf{x} + \mathbf{f}(\mathbf{x})$ is globally asymptotically stable. When $\tau = -4$ the system is point dissipative since $\mathbf{x}(A - 4BB^T)\mathbf{x} < 0$ for all nontrival zeros of $\mathbf{f}$.

Five trajectories of the control problem

$$\frac{d\mathbf{x}}{dt} = \left(\begin{bmatrix} 3 & 1 & -4 \\ 2 & -1 & 0 \\ -4 & 0 & 1 \end{bmatrix} + \tau \begin{bmatrix} 1 \\ 0 \\ -1 \end{bmatrix} \begin{bmatrix} 1 & 0 & -1 \end{bmatrix}\right)\mathbf{x} + \begin{bmatrix} 0 \\ -xz \\ xy \end{bmatrix}$$

are shown in figure 1. The initial point of all of the trajectories is(- 10, 10, 10). One trajectory is when there is no control, that is, $\tau = 0$. This is the long trajectory that is at (- 329, -4, 7) when t = 1.2. This trajectory seems to be moving right along at that time. The other trajectories are when the control parameter $\tau = -4, -5, -6$, and -7. These trajectories seem to reach a limit point of the controlled system, namely, (- 1.7, - 1.7, 1.0),

(- 0.5, -1.0, 0.0), (0.0, 0.0, 0.0) and (0.0, 0.0, 0.0), respectively. These trajectories have slowed down and terminate when $t = 600$. The first trajectory uses 74723 calculations while the last four trajectories only use about 7500 each.. The adaptive numerical method used to calculate the the trajctories takes larger time steps as the trajectory slows down.

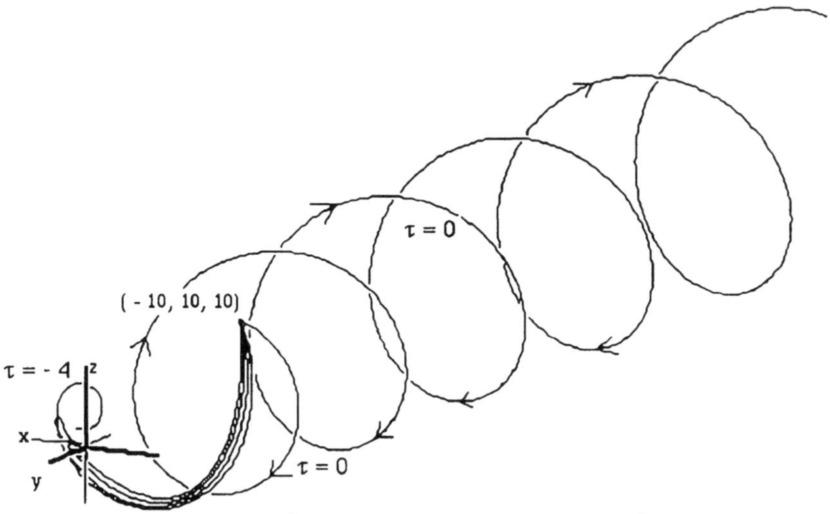

$$\frac{d\mathbf{x}}{dt} = \left(\begin{bmatrix} 3 & 1 & -4 \\ 2 & -1 & 0 \\ -4 & 0 & 1 \end{bmatrix} + \tau \begin{bmatrix} 1 \\ 0 \\ -1 \end{bmatrix} \begin{bmatrix} 1 & 0 & -1 \end{bmatrix} \right) \mathbf{x} + \begin{bmatrix} 0 \\ -xz \\ xy \end{bmatrix}$$

Figure 1

Theorem 3. Consider the system $\mathbf{M}\mathbf{x}' = N + (A + BK)\mathbf{x} + \mathbf{f}(\mathbf{x})$ with A, B and **f** fixed. If Z_f contains an (n - 1)-dimensional subspace of R^n and either

1) $Z_f \cap B^\perp = \{0\}$ or

2) $Z_f \cap B^\perp \neq \{0\}$ and $\mathbf{u}^T A \mathbf{u} < 0$ for all nontrivial **u** in $Z_f \cap B^\perp$

then there exists a matrix K such that the system is point dissipative.

Example for Theorem 3. Consider the control system $\mathbf{x}' = A\mathbf{x} + \mathbf{f}(\mathbf{x}) + B\mathbf{u}$, where

$$A = \begin{bmatrix} -1 & 1 & 3 \\ 2 & 1 & 2 \\ 3 & 1 & 1 \end{bmatrix}, B = \begin{bmatrix} 0 \\ 1 \\ 1 \end{bmatrix} \text{ and } \mathbf{f}(\mathbf{x}) = \begin{bmatrix} 0 \\ -x_1 x_3 \\ x_1 x_2 \end{bmatrix}$$

Note that A is not a definite matrix, $[1 \quad 0 \quad 0]A\begin{bmatrix}1\\0\\0\end{bmatrix} = -1 < 0$ and $[0 \quad 1 \quad 1]A\begin{bmatrix}0\\1\\1\end{bmatrix} = 5 > 0$.

$B^{\perp} = S\left\{\begin{bmatrix}1\\0\\0\end{bmatrix}, \begin{bmatrix}0\\-1\\1\end{bmatrix}\right\}$. If **u** is in $B^{\perp}$ then $\mathbf{u} = \begin{bmatrix}u_1\\-u_2\\u_2\end{bmatrix}$. $Z_f = S\left\{\begin{bmatrix}0\\1\\1\end{bmatrix}, \begin{bmatrix}0\\0\\1\end{bmatrix}\right\} \cup S\left\{\begin{bmatrix}1\\0\\0\end{bmatrix}\right\}$

$Z_f = S\left\{\begin{bmatrix}0\\1\\1\end{bmatrix}, \begin{bmatrix}0\\0\\1\end{bmatrix}\right\} \cup S\left\{\begin{bmatrix}1\\0\\0\end{bmatrix}\right\}$. If **u** is in Z_f then either $\mathbf{u} = \begin{bmatrix}u_1\\0\\0\end{bmatrix}$ or $\mathbf{u} = \begin{bmatrix}0\\u_2\\u_3\end{bmatrix}$. Thus

$$Z_f \cap B^{\perp} = \left\{\begin{bmatrix}u_1\\0\\0\end{bmatrix}, \begin{bmatrix}0\\-u_2\\u_2\end{bmatrix} \mid u_1 \neq 0, u_2 \neq 0\right\}$$

Now $[u_1 \quad 0 \quad 0]A\begin{bmatrix}u_1\\0\\0\end{bmatrix} = -u_1^2 < 0$ and $[0 \quad u_2 \quad u_2]A\begin{bmatrix}0\\-u_2\\u_2\end{bmatrix} = u_2^2 - 3u_2^2 + u_2^2 = -u_2^2 < 0$, i.e.,

if **u** is a nontrivial element of $Z_f \cap B_{\perp}$ then $\mathbf{u}^T A\mathbf{u} < 0$. Since $\begin{bmatrix}2\\-1\\1\end{bmatrix}$ is in $B^{\perp}$ and

$\begin{bmatrix}2\\-1\\1\end{bmatrix}$ is in $B^{\perp}$ and $[2 \quad -1 \quad 1]A\begin{bmatrix}2\\-1\\1\end{bmatrix} = 1 > 0$ Theorem 2 does not apply. Again

$\begin{bmatrix} 0 \\ 1 \\ 1 \end{bmatrix}$ is in Z_f and $[0 \quad 1 \quad 1]A\begin{bmatrix} 0 \\ 1 \\ 1 \end{bmatrix} = 5 > 0$ and so Theorem 1 does not apply. If we choose K

$= -2B^T$ then $A + BK = A - 2BB^T = \begin{bmatrix} -1 & 1 & 3 \\ 2 & -1 & 0 \\ 3 & -1 & -1 \end{bmatrix}$ Note that $A - 2BB^T$ is not negative

definite, but $[u_1 \quad 0 \quad 0](A - 2BB^T)\begin{bmatrix} u_1 \\ 0 \\ 0 \end{bmatrix} = -u_1^2 < 0$ and $[0 \quad -u_2 \quad u_2](A - 2BB^T)\begin{bmatrix} 0 \\ -u_2 \\ u_2 \end{bmatrix}$

$= -u_2^2 < 0$. Therefore if $\mathbf{u}$ is in Z_f then $\mathbf{u}^T(A - 2BB^T)\mathbf{u} < 0$. Hence $\mathbf{x}' = (A - 2BB^T)\mathbf{x} + \mathbf{f}(\mathbf{x})$ is point dissipative. In figure 2 two trjectories are given. One when the system is uncontroled, $\tau = 0$, and one when the system is controled, $\tau = -2$. For both trajectories (- 2, - 2, 2) is the initial point.

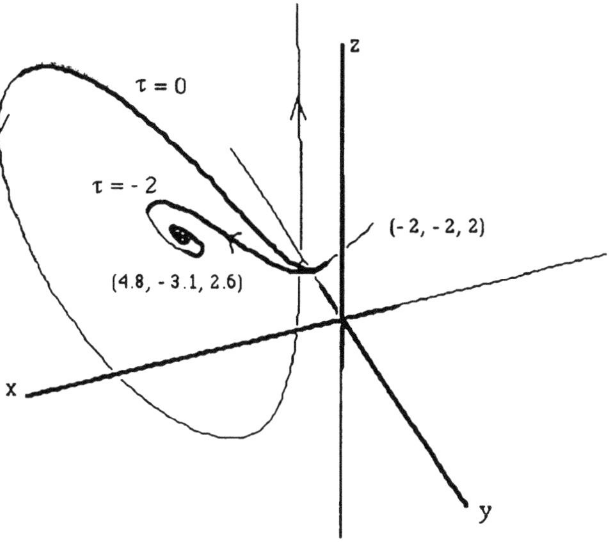

$$\frac{d\mathbf{x}}{dt} = \left(\begin{bmatrix} -1 & 1 & 3 \\ 2 & 1 & 2 \\ 3 & 1 & 1 \end{bmatrix} + \tau \begin{bmatrix} 0 \\ 1 \\ 1 \end{bmatrix} [0 \quad 1 \quad 1] \right) \mathbf{x} + \begin{bmatrix} 0 \\ -xz \\ xy \end{bmatrix}$$

Figure 2

4. Extension of previous results

The sufficient condition for a quadratic dynamical system to be point dissipative discussed above uses a relation between the quadratic and linear parts of the system when **f** is conservative. The following lemma allows us to extend the condition to the case where there exists a positive definite matrix S such that SM = MS and **Sf** is conservative. See [4].

Lemma 6. Let

$$\mathbf{Mx}' = \mathrm{N} + \mathrm{A}\mathbf{x} + \mathbf{f}(\mathbf{x}) \qquad (2.1)$$

be a quadratic dynamical system for which there exists a positive definite matrix S such that SM = MS and **Sf** is conservative. Then there exists matrix H for which the change of variables $\mathbf{y} = \mathrm{H}\mathbf{x}$ transforms the dynamical system (2.1) into $(\mathrm{HMH}^{-1})\mathbf{y}' = \mathrm{HN} + \mathrm{B}\mathbf{y} + \mathbf{g}(\mathbf{y})$ which has a conservative quadratic term. Furthermore, $\mathbf{x}^T\mathrm{SA}\mathbf{x} = \mathbf{y}^T\mathrm{B}\mathbf{y}$.

When **Sf** is conservative then we say that a vector α is admissible for the dynamical system $\mathbf{Mx}' = \mathrm{N} + \mathrm{A}\mathbf{x} + \mathbf{f}(\mathbf{x})$ if $-\mathbf{x}^T\mathrm{SA}\mathbf{x} + \alpha^T \mathrm{H}\, \mathbf{f}(\mathbf{x})$ is a positive definite function where $\mathrm{S} = \mathrm{H}^T\mathrm{H}$.

When **f** is conservative then the condition for an admissible α reduces to $-\mathbf{x}^T\mathrm{A}\mathbf{x} + \alpha^T\mathbf{f}(\mathbf{x})$ is positive definite. The proofs of the theorems when the quadratic part of the system is conservative entail demonstrating the existence of an admissible α for the system.

These results can be restated as follows:

Theorem 4. A quadratic dynamical system $\mathbf{Mx}' = \mathrm{N} + \mathrm{A}\mathbf{x} + \mathbf{f}(\mathbf{x})$ is point dissipative when there exists a positive definite matrix S such that SM = MS and **Sf** is conservative and there exists an admissible α.

Another direction of generalizing the past results is to consider nonlinear dynamical systems which have nonquadratic nonlinear terms as well as quadratic terms. Relative to the nonlinear terms there again must exit a positive definite matrix S such that the nonlinear terms premultiplied by S are conservative. See [4].

Theorem 5. When there exists a positive definite matrix S such that SM = MS, **Sg** and the quadratic function **Sf** are conservative, and

Condition (A) For some admissible α for $\mathbf{Mx}' = \mathrm{N} + \mathrm{A}\mathbf{x} + \mathbf{f}(\mathrm{x})$ there exists an ordered triple of numbers $(\varepsilon, \mathrm{C}, \mathrm{M})$ such that $-\alpha^T\mathrm{H}\mathbf{g}(\mathrm{x}) \leq \mathrm{C}\, \|\mathrm{x}\|^{2-\varepsilon}$ for all $\mathbf{x}$ with $\|\mathrm{x}\| \geq \mathrm{M}$.

then $\mathbf{Mx}' = \mathrm{N} + \mathrm{A}\mathbf{x} + \mathbf{f}(\mathrm{x}) + \mathbf{g}(\mathbf{x})$ is point dissipative.

Note that condition (A) can be replaced by either of the stronger conditions (B) or (C).

Condition (B) There is an admissible α for $\mathbf{Mx}' = \mathrm{N} + \mathrm{A}\mathbf{x} + \mathbf{f}(\mathrm{x})$ and $\| \mathbf{g} \| = o \| \mathbf{x} \|^2$.

Condition (C) There is an admissible α for $\mathbf{Mx}' = \mathrm{N} + \mathrm{A}\mathbf{x} + \mathbf{f}(\mathrm{x})$ and **g** is bounded.

Theorem 6. A quadratic dynamical system $Mx' = N + Ax + f(x)$, A a matrix and f a quadratic function, is point dissipative when

(1) there exists a positive definite matrix S such that $SM = MS$ and Sf is conservative,

(2) the zeros of f contains an (n - 1)-dimensional hyperplane,

and

(3) $z^TSAz < 0$ for any z which is a nontrivial zero of f.

This result can be generalized by adding to the differential equation any conservative function $g(x)$ whose growth is restricted. The corollary states this condition.

Corollary. Let g and the quadratic function f be conservative. If Z(f) contains an (n - 1)-hyperplane and $x \neq 0$ in Z(f) implies $x^TA\,x < 0$ and

Condition (A) For some admissible α there exists an ordered triple of numbers (ε, C, M) such that $-\alpha^Tg(x) \leq \|x\|^{2-\varepsilon}$ for all x with $\|x\| \geq M$.

then $Mx' = N + Ax + f(x) + g(x)$ is point dissipative.

Note that condition (A) can be replaced by either of the stronger conditions (B) or (C).

Condition (B) There is an admissible α for $Mx' = N + Ax + f(x)$ and $\| g \| = o \| x \|^2$.

Condition (C) There is an admissible α for $Mx' = N + Ax + f(x)$ and g is bounded.

5. Conclusions

Quadratic and "almost" quadratic nonlinear systems can exhibit a wide range of qualitative behavior. Even the subclass of such systems with compact attractors contains systems with point attractors, limit cycles and strange attractors [6]. Linear feedback control problems with system objectives of steering to desired limit points or of minimizing the diameter of a compact attractor have yet to be formulated and solved. This paper represents only a first step, i.e., using linear feedback to produce a controlled system with a compact attractor.

References

1. A. K. Bose and J. A. Reneke, Sufficient conditions for two-dimensional point dissipative systems, International Journal of Mathematics and Mathematical Sciences, 12(1989), 693-696.

2. A. K. Bose, A. S. Cover, and J. A. Reneke, On Point-dissipative Systems of Differential Equations with Quadratic Nonlinearity, International Journal of Mathematics and Mathematical Sciences, 14(1989), 99-110.

3. A. K. Bose, A. S. Cover, and J. A. Reneke, A class of point dissipative n-dimensional nonlinear dynamical systems, Proceedings of the Twenty-Fourth Southeastern Symposium on System Theory, 1992, 2-6.

4. A. K. Bose, A. S. Cover, and J. A. Reneke, On point dissipative n-dimensional systems of differential equations with quadratic nonlinearity, International Journal of Mathematics and Mathematical Sciences, accepted.

5. E. Lorenz, Deterministic non-periodic flow, Journal of Atmospheric Science, 20(1963), 130-14.

6. C. Sparrow, The Lorenz Equations: Bifurcations, Chaos, and Strange Attractors, Springer-Verlag, New York, 1982.

Third International Colloquium on Differential Equations pp. 23-30 (1993)
D. Bainov and V. Covachev (Eds)

OSCILLATORY PROPERTIES FOR DAMPED HYPERBOLIC EQUATIONS WITH DEVIATING ARGUMENTS

D.D. BAINOV[1] and CUI BAOTONG[2]
[1]Academy of Medicine, P. O. Box 45, Sofia 1504, BULGARIA
[2]Dept. of Math., Binzhou Normal College, Shandong 256604, P.R. CHINA

Abstract. Sufficient conditions for oscillation of the solutions of damped hyperbolic equations with deviating arguments are found.

Keywords: oscillatory, damped hyperbolic equations, deviating argument.

1. INTRODUCTION

Nowadays one can observe an expanding interest toward the study of oscillations for partial differential equations with deviating arguments. We refer the reader to Mishev and Bainov [1, 2], Yoshida [3], Georgiou & Kreith [4] and Cui [5, 6]. However, the oscillations of solutions of damped partial differential equations with deviating arguments have not been studied till now.

This paper gives some sufficient conditions for oscillation of solutions of the damped hyperbolic equation with deviating arguments

$$u_{tt} - b(t)\Delta u - \sum_{j=1}^{k} b_j(t)\Delta u(x,t-r_j) + p(t)u + \sum_{i=1}^{m} q_i(x,t)u(x,t-\tau_i) + a(t)u_t = 0, \quad (x, t) \in \Omega \times (0, \infty), \tag{1}$$

where Ω is a bounded domain in $\mathbb{R}^n$ with piecewise smooth boundary $\partial\Omega$, $u = u(x, t)$ and Δ is the Laplacian in the Euclidean n-space $\mathbb{R}^n$.

We assume throughout this paper that

(i) $p(t) \in C([0,\infty); [0,\infty))$, $q_i \in C(\bar{G}; [0,\infty))$, $i = 1, 2, \dots, m$, $G = \Omega \times (0, \infty)$ and b_j, b, $a \in C([0,\infty); [0,\infty))$, $j = 1, 2, \dots, k$.

(ii) $\tau_i > 0$ and $r_j > 0$ are constants and $q_i(t) = \min_{x\in\bar{\Omega}} q_i(x, t)$ $(i = 1, 2, \ldots, m)$ are not identically zero on any ray $[t, \infty)$.

Consider two kinds of boundary conditions

$$(B_1) \qquad \frac{\partial u}{\partial N} + r(x,t)u = 0 \quad \text{on } \partial\Omega \times [0, \infty),$$

where N is the unit exterior normal vector to $\partial\Omega$ and $r(x,t)$ is a non-negative continuous function on $\partial\Omega \times [0, \infty)$, and

$$(B_2) \qquad u = 0 \quad \text{on } \partial\Omega \times [0, \infty).$$

Our objective is to present conditions which imply that every solution $u(x, t)$ of (1) satisfying a certain boundary condition is oscillatory in $\Omega \times [0, \infty)$ in the sense that u has a zero in $\Omega \times [t, \infty)$ for any $t > 0$.

2. OSCILLATION FOR THE PROBLEM (1), (B_1)

Theorem 1. If

$$(C_1) \qquad \int^{\infty} [p(t) - \frac{a^2(t)}{4}]dt = \infty,$$

(C_2) there exist $\alpha \in (1, \infty)$ and $\beta \in [0, 1)$ such that

$$\limsup_{t\to\infty} \frac{1}{t^{\alpha}} \int_{t_0}^{t} \{(t - y)^{\alpha}y^{\beta}[p(y) + \sum_{i=1}^{m} q_i(y)]$$

$$- \frac{1}{4} [(t-y)ya(y) + \alpha y - \beta(t-y)]^2(t-y)^{\alpha-2}y^{\beta-2}\}dy = \infty,$$

then every solution of the problem (1), (B_1) oscillates in G.

Proof. Suppose to the contrary that there is a solution $u(x, t)$ of the problem (1), (B_1) which has no zero in $\Omega \times [t_0, \infty)$ for some $t_0 > 0$. Let $u > 0$ in $\Omega \times [t_0, \infty)$. Then there exists $t_1 \geq t_0$ such that $u(x, t - \tau_i) > 0$ and $u(x, t - r_j) > 0$ $(i = 1, 2, \ldots, m,$ and $j = 1, 2, \ldots, k)$ for any $(x, t) \in \Omega \times [t_1, \infty)$.

Integrating (1) over Ω, we have that

$$(2) \quad \frac{d^2}{dt^2}\left(\int_{\Omega} udx\right) - b(t)\int_{\Omega}\Delta udx - \sum_{j=1}^{k} b_j(t)\int_{\Omega}\Delta u(x,t-r_j)dx + p(t)\int_{\Omega}udx$$

$$+ \sum_{i=1}^{m} \int_{\Omega} q_i(x,t)u(x,t-\tau_i)dx + a(t)\frac{d}{dt}\left(\int_{\Omega} udx\right) = 0, \quad t \geq t_1 .$$

It follows from Green's formula that, for $t \geq t_1$,

$$(3) \qquad \int_{\Omega}\Delta udx = \int_{\partial\Omega} \frac{\partial u}{\partial N} dS = - \int_{\partial\Omega} r(x,t)u(x,t)dS \leq 0, \text{ and}$$

$$\int_\Omega \Delta u(x,t-r_j)dx = -\int_{\partial\Omega} r(x,t-r_j)u(x,t-r_j)dS \le 0,\ j = 1,2,\dots,k,$$

(4) $\int_\Omega q_i(x,t)u(x,t-\tau_i)dx \ge q_i(t)\int_\Omega u(x,t-\tau_i)dx,\ i = 1, 2, \dots, m.$

Thus, by (2) ~ (4) we obtain that

(5) $v''(t) + p(t)v(t) + \sum_{i=1}^m q_i(t)v(t-\tau_i) + a(t)v'(t) \le 0,\ t \ge t_1,$

where $v(t) = \int_\Omega u(x,t)dx$.

Set $w(t) = v'(t)/v(t)$ for $t \ge t_1$, then from (5) we have that

(6) $w'(t) + a(t)w(t) + w^2(t) + p(t) + \sum_{i=1}^m q_i(t)\frac{v(t-\tau_i)}{v(t)} \le 0,\ t \ge t_1.$

Then by (6)

$$w'(t) + [w(t) + \frac{a(t)}{2}]^2 - \frac{a^2(t)}{4} + p(t) \le 0,\quad t \ge t_1, \text{ and}$$

(7) $w'(t) + \left(p(t) - \frac{a^2(t)}{4}\right) \le 0,\quad t \ge t_1.$

Integrating (7) from t_1 to $t \ge t_1$, we have

$$w(t) - w(t_1) \le -\int_{t_1}^t [p(t) - \frac{a^2(t)}{4}]dt,\quad t \ge t_1.$$

It follows from (C_1) that there exists $t_2 \ge t_1$ such that $w(t) < 0$ and $v'(t) < 0$ for $t \ge t_2$. Therefore, we have that $\frac{v(t-\tau_i)}{v(t)} > 1$ for $t \ge t_2$ and all i. So (6) implies that

(8) $w'(t) + a(t)w(t) + w^2(t) + \left(p(t) + \sum_{i=1}^m q_i(t)\right) \le 0,\quad t \ge t_2,$

and consequently, for $t > y \ge t_2$,

$$\int_y^t (t-y)^\alpha y^\beta w'(y)dy + \int_y^t (t-y)^\alpha y^\beta a(y)w(y)dy + \int_y^t (t-y)^\alpha y^\beta w^2(y)dy$$
$$+ \int_y^t (t-y)^\alpha y^\beta \left(p(y) + \sum_{i=1}^m q_i(y)\right)dy \le 0.$$

Since

$$\int_y^t (t-y)^\alpha y^\beta w'(y)dy = \alpha\int_y^t (t-y)^{\alpha-1}y^\beta w(y)dy,$$

we have that

$$\int_y^t (t-y)^\alpha y^\beta [p(y) + \sum_{i=1}^m q_i(y)]dy \le w(y)(t-y)^\alpha y^\beta - \int_y^t (t-y)^\alpha y^\beta w^2(y)dy$$
$$- \int_y^t [(t-y)a(y)y + \alpha y - \beta(t-y)](t-y)^{\alpha-1}y^{\beta-1}w(y)dy,$$

and hence

$$\int_y^t (t-y)^\alpha y^\beta [p(y) + \sum_{i=1}^m q_i(y)]dy \le w(y)(t-y)^\alpha y^\beta$$

$$- \int_y^t \{(t-y)^{\alpha/2} y^{\beta/2} w(y) + H(y,t)\}^2 dy + \int_y^t H^2(y,t)dy,$$

where $H = \frac{1}{2} [(t-y)ya(y) + \alpha y - \beta(t-y)](t-y)^{\alpha/2-1} y^{\beta/2-1}$. Thus, we obtain that

$$\int_y^t (t-y)^\alpha y^\beta \left(p(y) + \sum_{i=1}^m q_i(y)\right) dy - \int_y^t H^2(y,t)dy \le w(y)(t-y)^\alpha y^\beta, \quad t > y \ge t_2 . \tag{9}$$

Divide (9) by t^α and take the upper limit as $t \longrightarrow \infty$. Using (C_2), we can obtain a contradiction.

If $u(x,t) < 0$ for $(x,t) \in \Omega \times [t_0 , \infty)$, then the proof follows from the fact that $-u(x,t)$ is a positive solution of the problem (1),(B_1). This completes the proof of the theorem.

Corollary 2. Assume that (C_1) holds and there exist $\alpha \in (1, \infty)$ and $\beta \in [0, 1)$ such that

$$\limsup_{t \longrightarrow \infty} t^{-\alpha} \int_{t_0}^t (t-y)^\alpha y^\beta [p(y) + \sum_{i=1}^m q_i(y)]dy = \infty,$$

$$\limsup_{t \longrightarrow \infty} t^{-\alpha} \int_{t_0}^t [(t-y)ya(y) + \alpha y - \beta(t-y)]^2 (t-y)^{\alpha-2} y^{\beta-2} dy < \infty,$$

then every solution of the problem (1), (B_1) oscillates in G.

From the proof of Theorem 1 it is easy to obtain the following theorem.

Theorem 3. If the inequality

$$v''(t) + p(t)v(t) + \sum_{i=1}^m q_i(t)v(t-\tau_i) + a(t)v'(t) \le 0$$

has no ultimately positive solutions, then every solution of the problem (1), (B_1) oscillates in G.

Theorem 4. If

(a) $\displaystyle \lim_{t \longrightarrow \infty} \int_{t_0}^t \exp\left(- \int_{t_0}^s a(y)dy\right) ds = \infty$, for every $t_0 \ge 0$,

(b) there exist $\alpha \in (1, \infty)$, $\beta \in [0, 1)$ and $\varepsilon \in (0, 1)$ such that

$$\limsup_{t \to \infty} \frac{1}{t^{\alpha}} \int_{t_0}^{t} \left\{ (t-y)^{\alpha} y^{\beta} \left[p(y) + \varepsilon \sum_{i=1}^{m} \frac{y-\tau_i}{y} q_i(y) \right] - \frac{1}{4} [(t-y)ya(y) + \alpha y - \beta(t-y)]^2 (t-y)^{\alpha-2} y^{\beta-2} \right\} dy = \infty,$$

then every solution of the problem (1), (B_1) oscillates in G.

<u>Proof.</u> Let $u(x,t)$ be a nonoscillatory solution of the problem (1), (B_1). Without loss of generality, we have that (5) holds. From (a) it follows that there is $t_2 \ge t_1$ such that $v'(t) > 0$ for $t \ge t_2$ (see Lemma 5.1 of Kartsatos [7]). Then we have that $v''(t) \le 0$ for $t \ge t_2$ from (5) and (i). It follows that there exists $t_3 \ge t_2$ such that for $t \ge t_3$

$$v(t-\tau_i) \ge \varepsilon v(t)\frac{t-\tau_i}{t}, \quad i = 1, 2, \ldots, m, \tag{10}$$

(see Yan [8]). Then from (5) and (10) we have that

$$v''(t) + a(t)v'(t) + \left[p(t) + \varepsilon \sum_{i=1}^{m} \frac{t-\tau_i}{t} q_i(t)\right] \le 0, \quad t \ge t_3 . \tag{11}$$

Set $w(t) = v'(t)/v(t)$ for $t \ge t_3$. Then from (11)

$$w'(t) + a(t)w(t) + w^2(t) + \left[p(t) + \varepsilon \sum_{i=1}^{m} \frac{t-\tau_i}{t} q_i(t)\right] \le 0, \quad t \ge t_3.$$

The remains are similar to the proof of Theorem 1 and we omit it.

<u>Theorem 5.</u> If (a) holds and there exists $\varepsilon \in (0, 1)$ so that the inequality

$$H'(t) + (\varepsilon t p(t) + a(t))H(t) + \varepsilon \sum_{i=1}^{m} q_i(t)(t-\tau_i)H(t-\tau_i) \le 0$$

has no ultimately positive solutions, then every solution of problem (1), (B_1) oscillates in G.

<u>Proof.</u> As in the proof of Theorem 1, we can prove that (5) holds. From (a) it follows that there is $t_2 \ge t_1$ such that $v'(t) > 0$ for $t \ge t_2$ (see [7]). Then $v''(t) < 0$ for $t \ge t_2$ by (5) and (i). Thus we have that there exists $t_3 \ge t_2$ so that $v(t) \ge \varepsilon t v'(t)$ for $t \ge t_3$ (see

Lemma of Wei [9]). So we obtain from (5)

$$(12) \qquad v''(t)+\varepsilon tp(t)v'(t)+\varepsilon \sum_{i=1}^{m} q_i(t)(t-\tau_i)v'(t-\tau_i)+a(t)v'(t) \le 0$$

for $t \ge t_4 = t_3 + \max_{1\le i\le m} \{\tau_i\}$. Let $H(t) = v'(t)$ for $t \ge t_4$, then from (12) it follows that

$$(13) \qquad H'(t) + (\varepsilon tp(t)+a(t))H(t) + \varepsilon \sum_{i=1}^{m} q_i(t)(t-\tau_i)H(t-\tau_i) \le 0$$

for $t \ge t_4$ and $H(t) = v'(t)$ is an ultimately positive solution of the inequality (13), which contradicts the condition of the theorem.

3. OSCILLATION FOR THE PROBLEM (1), (B_2)

Consider the following Dirichlet problem in the domain Ω

$$(14) \qquad \begin{cases} \Delta w + \alpha w = 0 & \text{in } \Omega, \\ w\big|_{\partial\Omega} = 0, \end{cases}$$

where α = const. It is well known that the smallest eigenvalue α_0 of the problem (14) is positive and the corresponding eigenfunction $\Phi(x) > 0$ for $x \in \Omega$.

With each solution $u(x,t)$ of the problem (1),(B_2) we associate the function

$$(14) \qquad v(t) = \int_{\Omega} u(x,t)\Phi(x)dx, \qquad t \ge 0.$$

Theorem 6. If the delay inequality

$$v''(t) + [\alpha_0 b(t) + p(t)]v(t) + \alpha_0 \sum_{j=1}^{k} b_j(t)v(t-r_j) + \sum_{i=1}^{m} q_i(t)v(t-\tau_i) + a(t)v'(t) \le 0$$

has no ultimately positive solutions, then every solution of the problem (1), (B_2) oscillates in G.

Proof. Let $u(x, t)$ be a positive solution of the problem (1), (B_2) in G. Then there exists a number $t_0 > 0$ such that $u(x,t-\tau_i) > 0$ and $u(x,t-r_j) > 0$ ($i = 1, 2, \ldots, m$; $j = 1, 2, \ldots, k$) in $\Omega \times [t_0, \infty)$. Multiplying both sides of (1) by the eigenfunction $\Phi(x)$, and integrating with respect to x over Ω, we have that for $t \geq t_0$

(16) $$\frac{d^2}{dt^2}\left(\int_\Omega u(x,t)\Phi(x)dx\right) - b(t)\int_\Omega \Delta u\Phi dx - \sum_{j=1}^{k} b_j(t)\int_\Omega \Delta u(x,t-r_j)\Phi dx$$

$$+ p(t)\int_\Omega u\Phi dx + \sum_{i=1}^{m} \int_\Omega q_i(x,t)u(x,t-\tau_i)\Phi dx + a(t)\frac{d}{dt}\left(\int_\Omega u\Phi dx\right) \leq 0.$$

From the divergence theorem and (ii) it follows that

(17) $$\int_\Omega \Delta u\Phi dx = \int_{\partial\Omega}\left(\frac{\partial u}{\partial N}\Phi + u\frac{\partial\Phi}{\partial N}\right)dS + \int_\Omega u\Delta\Phi dx$$

$$= \int_\Omega u\Delta\Phi dx = -\alpha_0 \int_\Omega u\Phi dx;$$

(18) $$\int_\Omega q_i(x,t)u(x,t-\tau_i)\Phi dx \geq q_i(t)\int_\Omega u(x,t-\tau_i)\Phi dx, \quad i = 1, 2, \ldots, m,$$

$$\int_\Omega \Delta u(x,t-r_j)\Phi dx = \int_{\partial\Omega}\left(\frac{\partial u(x,t-r_j)}{\partial N}\Phi - u(x,t-r_j)\frac{\partial\Phi}{\partial N}\right)dS + \int_\Omega u(x,t-r_j)\Delta\Phi dx$$

$$= \int_\Omega u(x,t-r_j)\Delta\Phi dx = -\alpha_0 \int_\Omega u(x,t-r_j)\Phi dx, \quad j = 1, 2, \ldots, k.$$

Combining (6), (17) with (18), we obtain

(19) $$v''(t)+[\alpha_0 b(t)+p(t)]v(t)+ \alpha_0 \sum_{j=1}^{k} b_j(t)v(t-r_j)+ \sum_{i=1}^{m} q_i(t)v(t-\tau_i)$$

$$+ a(t)v'(t) \leq 0, \quad t \geq t_0 .$$

Thus $v(t) = \int_\Omega u(x,t)\Phi(x)dx > 0$ for $t \geq t_0$ is a positive solution of the inequality (19), which contradicts the condition of the theorem.

Similarly to Theorems 1-5, using the inequality (19), we obtain the following theorems.

<u>Theorem 7.</u> If $\int_{t_0}^{\infty} [\alpha_0 b(t) + p(t) - \frac{a^2(t)}{4}]dt = \infty$ and there exist $\alpha \in (1, \infty)$ and $\beta \in [0, 1)$ such that

$$\limsup_{t\to\infty} \frac{1}{t^\alpha}\int_{t_0}^{t}\left\{(t-y)^\alpha y^\beta\left(\alpha_0 b(y) + p(y) + \alpha_0 \sum_{j=1}^{k} b_j(y) + \sum_{i=1}^{m} q_i(y)\right)\right.$$

$$\left. - \frac{1}{4}[(t-y)ya(y) + \alpha y - \beta(t-y)]^2(t-y)^{\alpha-2}y^{\beta-2}\right\} dy = \infty,$$

then every solution of the problem (1), (B_2) oscillates in G.

<u>Theorem 8.</u> If

$$\lim_{t\to\infty} \int_{t_0}^{t} \exp\left(-\int_{t_0}^{s} a(y)dy\right)ds = \infty \quad \text{for every } t_0 \geq 0,$$

and there exist $\alpha \in (1, \infty)$, $\beta \in [0, 1)$ and $\varepsilon \in (0, 1)$ such that

$$\limsup_{t\to\infty} \frac{1}{t^{\alpha}} \int_{t_0}^{t} \left\{ (t-y)^{\alpha} y^{\beta} \left[\alpha_0 b(y) + p(y) + \alpha_0 \varepsilon \sum_{j=1}^{k} \frac{y-r_j}{y} b_j(y) + \varepsilon \sum_{i=1}^{m} \frac{y-\tau_i}{y} q_i(y) \right] - \frac{1}{4} [(t-y)ya(y) + \alpha y - \beta(t-y)]^2 (t-y)^{\alpha-2} y^{\beta-2} \right\} dy = \infty,$$

then every solution of the problem (1), (B_2) oscillates in G.

Theorem 9. If (a) (in Theorem 4) holds, and there exists $\varepsilon \in (0, 1)$ such that the inequality

$$H'(t) + [\varepsilon t(\alpha_0 b(t) + p(t)) + a(t)]H(t) + \alpha_0 \varepsilon \sum_{j=1}^{k} b_j(t)(t-r_j)H(t-r_j) + \varepsilon \sum_{i=1}^{m} q_i(t)(t-\tau_i)H(t-\tau_i) \le 0$$

has no ultimately positive solution, then every solution of the problem (1), (B_2) oscillates in G.

References

1. D.P. Mishev and D.D. Bainov, Appl. Math. Comput., 28, 97-111 (1988).
2. ___, Funk. Ekvac., 29, 213-218 (1986).
3. N. Yoshida, Bull. Austral. Math. Soc., 36, 289-294 (1987).
4. D. Georgiou and K. Kreith, J. Math. Anal. Appl., 107, 414-424 (1985).
5. Cui Baotang, Math. J. Toyama Univ., 14, 113-123 (1991).
6. ——, Oscillation Properties for Parabolic Equations of Neutral Type, Comment. Math. Univ. Carolinae, Vol. 33, No. 4 (in press).
7. A.K. Kartsatos, Stability of Dynamical Systems: Theory and Applications, Lecture Notes in Pure and Applied Mathematics, 28, Springer, New York (1977) pp.17-72.
8. Jurang Yan, J. Math. Anal. Appl., 122, 380-384 (1987).
9. Wei Junjie, Ann. Diff. Eqs., 4, 473-478 (1988).

Third International Colloquium on Differential Equations pp. 31-44 (1993)
D. Bainov and V. Covachev (Eds)

INITIAL VALUE PROBLEM FOR A SINGULARLY PERTURBED NONLINEAR IMPULSIVE SYSTEM IN THE CRITICAL CASE

V. COVACHEV
Institute of Mathematics, Bulgarian Academy of Science, Sofia, BULGARIA

Abstract. An asymptotic method for solving the initial value problem is justified for a nonlinear singularly perturbed impulsive system of differential equations in the "critical" case.

Keywords: critical case, initial value, impulsive, singularly perturbed.

1. INTRODUCTION

The interest in differential equations with impulse effect steadily increases in relation to the investigation of mathematical models of real processes and phenomena which at certain moments of their evolution undergo rapid changes.

The initial value problem in the so called noncritical case was investigated in [1]. In the present paper an asymptotic method for solving the initial value problem is justified for a nonlinear singularly perturbed impulsive system of differential equations in the so called critical case, that is when the linearized degenerate system has an eigenvalue zero of constant multiplicity.

The plan of the paper is as follows. In § 2 the necessary assertions from [2] concerning the existence of a stable manifold in the critical case are given. In § 3 the problem is formulated. In § 4 the construction of an asymptotic solution of this problem is carried out, and in § 5 the asymptotic method is justified, i.e. the existence of an exact solution of the problem considered is proved and an estimate of the difference between the exact and the asymptotic solution is obtained.

2. AUXILIARY ASSERTIONS

Consider the equation

(1) $$\dot{z} = F(z, \alpha),$$

where $z \in \mathbb{R}^m$, α is a parameter, F is continuously differentiable with respect to z, $F(0,\alpha) = 0$, $F_z(0,\alpha)$ has an eigenvalue 0 of multiplicity k (independent of α) and $m - k$ eigenvalues with negative real parts. Suppose, moreover, that for any value of the parameter α there are k linearly independent null eigenvectors,and denote by $\varphi(\alpha)$ an $(m \times k)$-matrix of rank k (whose columns are the linearly independent null eigenvectors) such that

$$F_z(0, \alpha)\varphi(\alpha) = 0.$$

Henceforth we shall often suppress the dependence on the parameter α, and denote by A the matrix $F_z(0, \alpha)$.

The stationary point $z = 0$ is in general not asymptotically stable, i.e. an arbitrary solution of (1) with initial condition arbitrarily close to 0 does not have to tend to 0 as $t \longrightarrow \infty$. However, if the initial condition is chosen in an appropriate way, then the solution will exponentially tend to 0 as $t \longrightarrow \infty$. More precisely, the following assertion is valid.

Lemma 1. In a sufficiently small neighbourhood of the point $z = 0$ there exists an $(m-k)$-dimensional manifold ω and positive constants γ and σ such that if $z(0) \in \omega$, then for $t \geq 0$ the solution $z(t)$ satisfies the inequality

(2) $$|z(t)| \leq \gamma e^{-\sigma t}.$$

The manifold ω is constructed by the method of successive approximations, representing the right-hand side of (1) as a sum of a linear term Az and a nonlinearity $G(z) \equiv F(z) - Az$:

(3) $$\dot{z} = Az + G(z).$$

The function $G(z)$ obviously enjoys the following two properties:

1. $G(0) = 0$.
2. For any $\varepsilon > 0$ there exists $\delta > 0$ such that for $|z_1| \leq \delta$, $|z_2| \leq \delta$ we have

$$|G(z_1) - G(z_2)| \le \varepsilon |z_1 - z_2|.$$

The matrix A has a k-tuple eigenvalue 0 and m - k eigenvalues with negative real parts. Then there exists a matrix B, as smooth with respect to the parameter α as A is, which reduces A to a block-diagonal form

(4) $$B^{-1}AB = \begin{pmatrix} C & 0 \\ 0 & 0 \end{pmatrix},$$

where the $(m-k) \times (m-k)$ -matrix C has the above mentioned eigenvalues $\lambda_1, \dots, \lambda_{m-k}$.

If for (1) we consider the linearized system, i.e. we set in (3) $G(z) = 0$, and denote $z = \begin{pmatrix} x \\ y \end{pmatrix}$, $x \in \mathbb{R}^{m-k}$, $y \in \mathbb{R}^k$, $B = \begin{pmatrix} B_{11} & B_{12} \\ B_{21} & B_{22} \end{pmatrix}$, where B_{ij}, i, j = 1, 2, are blocks of appropriate dimensions, in a linear approximation the manifold ω can be represented in the form

(5) $$y = B_{21}B_{11}^{-1}x$$

provided that the matrix B_{11} is nonsingular.

By Lemma 1 the manifold ω is constructed locally (in a neighbourhood of 0). Extending to the direction of $t < 0$ the trajectories starting on ω, we obtain an extended manifold Ω enjoying the same properties as ω, i.e. any trajectory $z(t)$ starting on Ω for $t = 0$ remains on Ω for all $t > 0$ and exponentially tends to 0 as $t \longrightarrow \infty$. Formula (5) suggests us to assume that in some domain $\mathcal{D}(x, \alpha)$ the manifold $\Omega = \Omega(\alpha)$ can be represented in the form

(6) $$y = P(x, \alpha),$$

where $P(x, \alpha)$ is a sufficiently smooth function on $\mathcal{D}(x, \alpha)$. (The parameter α will be again suppressed henceforth).

Denote $H(x) = P_x(x)$, $F(z) = F(x, y) = \begin{pmatrix} F_1(x, y) \\ F_2(x, y) \end{pmatrix}$, $F_1 \in \mathbb{R}^{m-k}$, $F_2 \in \mathbb{R}^k$, $F_{11} = \partial_x F_1$, $F_{12} = \partial_y F_1$, $F_{21} = \partial_x F_2$, $F_{22} = \partial_y F_2$.

Now consider a nonhomogeneous system of equations whose corresponding homogeneous system is the system in variations for (1):

(7) $$\dot{\Delta} = F_z(t)\Delta + \psi(t),$$

where $F_z(t) \equiv F_z(z(t))$, $z(t) \in \Omega$, $\psi(t)$ is a vector-valued function. As above, we denote $\Delta = \begin{pmatrix} \Delta_1 \\ \Delta_2 \end{pmatrix}$, $\psi = \begin{pmatrix} \psi_1 \\ \psi_2 \end{pmatrix}$.

Lemma 2. The change of variables

(8) $$\Delta_1 = \delta_1 , \ \Delta_2 = H(t)\delta_1 + \delta_2 ,$$

where $H(t) \equiv H(x(t))$, reduces system (7) to the form

(9) $$\begin{aligned} \dot{\delta}_1 &= a_{11}\delta_1 + a_{12}\delta_2 + \psi_1 , \\ \dot{\delta}_2 &= a_{22}\delta_2 + (\psi_2 - H\psi_1), \end{aligned}$$

where

(10) $$a_{11} = F_{11} + F_{12}H, \ a_{12} = F_{12} , \ a_{22} = F_{22} - HF_{12} .$$

In the above formulae the dependence on t is suppressed too.

Now suppose that in system (7) the nonhomogeneous part $\psi(t)$ is the matrix $F_z(t)\varphi$. Then the second equation in (9) becomes

(11) $$\dot{\delta}_2 = a_{22}\delta_2 + [F_z\varphi]_2 - H[F_z\varphi]_1 .$$

Obviously, for the given nonhomogeneous part there is a particular solution of (7) $\Delta = -\varphi$ (independent of t, but possibly depending on the parameter α). From (8) it follows that the corresponding particular solution of equation (11) is $\delta_2 = H(t)\varphi_1 - \varphi_2 \equiv R(t)$. This proves the following

Lemma 3. The matrix $\delta_2 = R(t)$ is a solution of equation (11) with initial condition

$$\delta_2(0) = H(0)\varphi_1 - \varphi_2 = R(0).$$

Next we shall give some relations about the matrices $a_{11}(t)$ and $a_{22}(t)$ in (9). Denote for the sake of brevity by $H(\infty)$ the limit of $H(t) = H(x(t))$ as $t \longrightarrow \infty$, and analogously for other matrix-valued functions. Then $a_{11}(\infty) = B_{11}CB_{11}^{-1}$. Thus the eigenvalues of $a_{11}(\infty)$ are the eigenvalues of the matrix C, i.e. those eigenvalues of $F_z(0)$ which have negative real parts, while $a_{22}(\infty) = 0$. Moreover,

$$|a_{22}(t)| \le ce^{-\kappa t}, \ t \ge 0.$$

Lemma 4. The fundamental matrix $\Psi(t)$ ($\Psi(0) = E_k$) of the system $\dot{\delta}_2 = a_{22}(t)\delta_2$ satisfies the conditions:

1. There exists $\lim_{t \longrightarrow \infty} \Psi(t) = \Psi(\infty)$.
2. $\det \Psi(\infty) \ne 0$.
3. $|\Psi(t) - \Psi(\infty)| \le ce^{-\kappa t}$.

Recall that $R(t) = H(t)\varphi_1 - \varphi_2$. This matrix appeared in Lemma 3.

Lemma 5. det $R(\infty) \neq 0$.

3. STATEMENT OF THE PROBLEM

Consider the initial value problem

(12) $\varepsilon\dot{z} = f(t, z, \varepsilon), \quad t \in [0, T], \quad t \neq t_i$,

(13) $\Delta z(t_i) = I_i(z(t_i)), \quad i = \overline{1, p}$,

(14) $z(0, \varepsilon) = z_0$,

where $z \in \mathbb{R}^m$, $0 = t_0 < t_1 < \dots < t_p < t_{p+1} = T$, $I_i: \Omega \longrightarrow \mathbb{R}^m$, $i = \overline{1, p}$, Ω is a domain in $\mathbb{R}^m$, $\Delta z(t_i) = z(t_i+0) - z(t_i)$, the jump functions $I_i(z)$ and the initial condition z_0 are assumed independent of ε, for the sake of simplicity.

Assume that the following conditions hold:

H1. For some $n \geq 2$ the vector-valued function $f \in C^{n+2}(G, \mathbb{R}^m)$, where $G = [0, T] \times \Omega \times [0, \varepsilon_0]$.

H2. $I_i \in C^{n+2}(\Omega, \mathbb{R}^m)$.

The further assumptions will be made in the course of the exposition.

First consider the degenerate system

(15) $f(t, \bar{z}, 0) = 0.$

H3. For each $t \in [0, T]$ system (15) has a family of solutions of the form

$$\bar{z} = \varphi(t, \alpha_1, \dots, \alpha_k) \equiv \varphi(t, \alpha),$$

where $\alpha = (\alpha_1, \dots, \alpha_k) \in D \subset \mathbb{R}^k$, $0 < k \leq m$, are parameters, and

H3$'$. $\varphi \in C^{n+2}([0, T] \times D)$.

H3$''$. Rank $\varphi_\alpha(t, \alpha) = k$ for all $(t, \alpha) \in [0, T] \times D$, where $\varphi_\alpha = \partial_\alpha \varphi$ is the matrix of derivatives of φ with respect to α.

If we differentiate the identity

$$f(t, \varphi(t, \alpha), 0) = 0$$

with respect to α, we obtain

$$f_z(t, \varphi(t, \alpha), 0).\varphi_\alpha(t, \alpha) = 0, \quad (t, \alpha) \in [0, T] \times D.$$

This means that the matrix $F(t, \alpha) \equiv f_z(t, \varphi(t,\alpha), 0)$ has an eigenvalue $\lambda(t,\alpha) \equiv 0$, and the columns of $\varphi_\alpha(t, \alpha)$ are null eigenvectors. Since by assumption H3$''$ they are linearly independent, then the multiplicity of $\lambda \equiv 0$ is at least k.

H4. The multiplicity of the eigenvalue $\lambda \equiv 0$ is exactly k, and the remaining m-k eigenvalues $\lambda_j(t, \alpha)$, $j = \overline{1, m-k}$, of $F(t, \alpha)$ satisfy in $[0, T] \times D$ the inequalities

(16) $$\mathrm{Re}\, \lambda_j(t, \alpha) < 0.$$

Denote $F^{(i)}(\zeta, \alpha) = f(t_i, \varphi(t_i, \alpha) + \zeta, 0)$, $i = \overline{0, p}$, $A^{(i)}(\alpha) = F^{(i)}_\zeta(0, \alpha) \equiv f_z(t_i, \varphi(t_i, \alpha), 0)$. Obviously, each of the matrices $A^{(i)}(\alpha)$ has an eigenvalue $\lambda = 0$ of multiplicity k, and m - k eigenvalues $\lambda_j(t_i, \alpha)$, $j = \overline{1, m-k}$, satisfying (16).

Thus there exist matrices $B^{(i)}(\alpha)$, $i = \overline{0, p}$, as smooth as $A^{(i)}(\alpha)$ are, such that

$$(B^{(i)}(\alpha))^{-1}A^{(i)}(\alpha)B^{(i)}(\alpha) = \begin{pmatrix} C^{(i)}(\alpha) & 0 \\ 0 & 0 \end{pmatrix},$$

where the $(m-k) \times (m-k)$-matrices $C^{(i)}(\alpha)$ have eigenvalues $\lambda_j(t_i, \alpha)$, $j = \overline{1, m-k}$.

Henceforth we shall write $z = \begin{pmatrix} x \\ y \end{pmatrix}$, $x \in \mathbb{R}^{m-k}$, $y \in \mathbb{R}^k$,

$$B^{(i)}(\alpha) = \begin{pmatrix} B^{(i)}_{11}(\alpha) & B^{(i)}_{12}(\alpha) \\ B^{(i)}_{21}(\alpha) & B^{(i)}_{22}(\alpha) \end{pmatrix},$$

where the blocks $B^{(i)}_{11}$, $B^{(i)}_{12}$, $B^{(i)}_{21}$, $B^{(i)}_{22}$ are matrices of dimension respectively $(m-k) \times (m-k)$, $(m-k) \times k$, $k \times (m-k)$, $k \times k$.

H5. The matrices $B^{(i)}_{11}(\alpha)$, $i = \overline{0, p}$, are nonsingular for $\alpha \in D$.

We can apply Lemma 1 for any of the systems

(17_i) $$\dot{\zeta} = F^{(i)}(\zeta, \alpha),$$

$i = \overline{0, p}$. Thus we prove the existence of a stable manifold $\Omega^{(i)}(\alpha)$ of (17_i) such that a trajectory $\zeta^{(i)}(t)$ of (17_i) issuing from $\zeta^{(i)}(0) \in \Omega^{(i)}(\alpha)$ remains on $\Omega^{(i)}(\alpha)$ and exponentially tends to 0 as $t \longrightarrow +\infty$. Moreover, let us suppose that

H6. The manifolds $\Omega^{(i)}(\alpha)$ can be represented in the form

(18) $$y = P^{(i)}(x, \alpha),$$

where $P^{(i)}(x,\alpha)$ is a sufficiently smooth function in some domain $D(x, \alpha)$.

Further on we can use all the results of § 2, and also the notation used there with superscript (i), when necessary.

It will be proved that problem (12)-(14) has a solution $z(t,\varepsilon) = (x(t, \varepsilon), y(t, \varepsilon))$, and an approximation $Z_n(t, \varepsilon)$ to this solution will be found such that

$$\|Z_n(., \varepsilon) - z(., \varepsilon)\| \le C\varepsilon^{n+1}.$$

This approximation will be constructed in the form

$$Z_n(t, \varepsilon) = Z_n^{(i)}(t, \varepsilon), \quad t \in (t_i , t_{i+1}], \quad i = \overline{0, p},$$

(19) $$Z_n^{(i)}(t, \varepsilon) = \sum_{\nu=0}^{n} \varepsilon^{\nu}[\bar{z}_\nu(t) + \mathfrak{z}_\nu^{(i)}(\tau_i)],$$

where $\bar{z}_\nu(t)$ is the solution of an appropriate lower order initial value problem, and $\mathfrak{z}_\nu^{(i)}(\tau_i)$ are the solutions of appropriate "boundary layer equations" which are also lower dimensional and represented in the dilated time scales

(20) $$\tau_i = (t-t_i)/\varepsilon, \quad t \in (t_i, t_{i+1}), \quad i=\overline{0, p}.$$

For the functions of τ_i we use the corresponding Gothic letters.

4. CONSTRUCTION OF THE ASYMPTOTIC SOLUTION

We shall seek for a formal asymptotic representation of the solution $z(t, \varepsilon)$ of problem (12)-(14) in the form

(21) $$z(t,\varepsilon) = \bar{z}(t,\varepsilon) + \mathfrak{z}^{(i)}(\tau_i,\varepsilon), \quad t_i < t < t_{i+1},$$

where

(22) $$\bar{z}(t, \varepsilon) \sim \sum_{\nu=0}^{\infty} \bar{z}_\nu(t)\varepsilon^{\nu}, \quad t \in [0, T],$$

(23) $$\mathfrak{z}^{(i)}(\tau_i , \varepsilon) \sim \sum_{\nu=0}^{\infty} \mathfrak{z}_\nu^{(i)} (\tau_i)\varepsilon^{\nu}, \quad i = \overline{0, p}.$$

The coefficients in the expansions (23) are called boundary layer (or just boundary) functions [2] and on them the following additional condition is imposed

(24) $$\mathfrak{z}_\nu^{(i)}(+\infty) = 0 \quad (i = \overline{0, p}).$$

Moreover, they are expected to decay exponentially as $\tau_i \longrightarrow +\infty$.

We should note that not all assumptions have been yet made.

We substitute the series (20) into (12)-(14). We use representations in the form of series in powers of ε as follows:

$$f(t, z(t, \varepsilon), \varepsilon) = \bar{f}(t, \varepsilon) + \mathfrak{f}^{(i)}(\tau_i , \varepsilon), \quad t \in (t_i , t_{i+1}],$$

$i = \overline{0, p}$, where $\bar{f}(t, \varepsilon) \equiv f(t, \bar{z}(t, \varepsilon), \varepsilon) \sim \sum_{\nu=0}^{\infty} \bar{f}_\nu(t)\varepsilon^\nu$,

$$f^{(i)}(\tau_i, \varepsilon) \equiv$$

$$f\left(t_i+\varepsilon\tau_i,\ \bar{z}(t_i+\varepsilon\tau_i,\varepsilon)+\mathfrak{z}^{(i)}(\tau_i,\varepsilon),\ \varepsilon\right) - f\left(t_i+\varepsilon\tau_i,\ \bar{z}(t_i+\varepsilon\tau_i,\varepsilon),\ \varepsilon\right)$$

$$\sim \sum_{\nu=0}^{\infty} f_\nu^{(i)}(\tau_i)\varepsilon^\nu,$$

$$\bar{f}_0(t) = \bar{F}(t) \equiv f(t, \bar{z}_0(t), 0),$$

$$\bar{f}_1(t) = \bar{F}_z(t)\bar{z}_1(t) + \bar{F}_\varepsilon(t)$$

$$\equiv f_z(t, \bar{z}_0(t), 0)\bar{z}_1(t) + f_\varepsilon(t, \bar{z}_0(t), 0),$$

$$\ldots, \bar{f}_\nu(t) = \bar{F}_z(t)\bar{z}_\nu(t) + h_\nu(t), \ldots$$

$h_\nu(t)$, $\nu = 2, 3, \ldots$, are expressed in terms of $\bar{z}_\mu(t)$, $\mu = \overline{0, \nu-1}$, and are linear in $\bar{z}_{\nu-1}(t)$ for $\nu \geq 3$;

$$f_0^{(i)}(\tau_i) = F^{(i)}(\tau_i) \equiv f(t_i\ , \bar{z}_0(t_i+0) + \mathfrak{z}_0^{(i)}(\tau_i), 0),$$

$$f_1^{(i)}(\tau_i) =$$

$$F_z^{(i)}(\tau_i)\mathfrak{z}_1^{(i)}(\tau_i) + [F_z^{(i)}(\tau_i)-\bar{F}_z(t_i)].[\bar{z}_1(t_i+0)+\tau_i\dot{\bar{z}}_0(t_i+0)]$$

$$+ [F_t^{(i)}(\tau_i) - \bar{F}_t(t_i)]\tau_i + F_\varepsilon^{(i)}(\tau_i) - \bar{F}_\varepsilon(t_i), \ldots$$

$$f_\nu^{(i)}(\tau_i) = F_z^{(i)}(\tau_i)\mathfrak{z}_\nu^{(i)}(\tau_i) + p_\nu^{(i)}(\tau_i), \ldots$$

Here

$$\bar{F}(t) \equiv f(t,\bar{z}_0(t+0),0),\ F^{(i)}(\tau_i) \equiv f(t_i,\ \bar{z}_0(t_i+0)+\mathfrak{z}_0^{(i)}(\tau_i),0),$$

and the notation $F_t^{(i)}(\tau_i),\ldots, \bar{F}_\varepsilon(t)$, etc. has an obvious sense, $p_\nu^{(i)}(\tau_i)$ are expressed in terms of $\bar{z}_\mu(t_i+0)$, $\mu = \overline{0, \nu}$, and their derivatives, and of $\mathfrak{z}_\mu^{(i)}(\tau_i)$, $\mu = \overline{0, \nu-1}$. Further on,

$$I_i(z(t_i\ , \varepsilon)) = I_i(z(t_i-0, \varepsilon)) = I_i\left(\bar{z}(t_i\ , \varepsilon) + \mathfrak{z}^{(i-1)}\left(\frac{t_i-t_{i-1}}{\varepsilon}\right)\right)$$

$$\approx I_i(\bar{z}(t_i\ , \varepsilon)) \sim I_i(\bar{z}_0(t_i)) + \varepsilon\partial_z I_i(\bar{z}_0(t_i))\bar{z}_1(t_i) + \ldots$$

$$+ \varepsilon^\nu[\partial_z I_i(\bar{z}_0(t_i))\bar{z}_\nu(t_i) + \alpha_{i\nu}] + \ldots$$

Here $\mathfrak{z}^{(i-1)}\left(\dfrac{t_i-t_{i-1}}{\varepsilon}\right)$ are neglected since they are expected to decay exponentially as $\varepsilon \longrightarrow 0$. They are taken into account in the proof of the convergence of the method. $\alpha_{i\nu}$ are expressed in terms of $\bar{z}_\mu(t_i)$, $\mu = \overline{0, \nu-1}$.

Now problem (12)-(14) takes the form

$$\varepsilon \sum_{\nu=0}^{\infty} \dot{\bar{z}}_\nu(t)\varepsilon^\nu + \sum_{\nu=0}^{\infty} \dot{\mathfrak{z}}_\nu^{(i)}(\tau_i)\varepsilon^\nu = \bar{f}(t, \varepsilon) + f^{(i)}(\tau_i, \varepsilon),$$

$$\sum_{\nu=0}^{\infty}\left(\bar{z}_\nu(t_i+0) + \mathfrak{z}_\nu^{(i)}(0)\right)\varepsilon^\nu$$

$$= \bar{z}_0(t_i)+I_i(\bar{z}_0(t_i)) + \varepsilon\left(E+\partial_z I_i(\bar{z}_0(t_i))\right)\bar{z}_1(t_i)$$

$$+ \sum_{\nu=2}^{\infty}\varepsilon^\nu\left\{\left(E + \partial_z I_i(\bar{x}_0(t_i))\right)\bar{z}_\nu(t_i) + \alpha_{i\nu}\right\},$$

$$\sum_{\nu=0}^{\infty}\left(\bar{z}_\nu(0) + \mathfrak{z}_\nu^{(0)}(0)\right)\varepsilon^\nu = z_0 .$$

In fact, the above expansions have sense only up to order n+1. We equate the coefficients at the like powers of ε, separately those depending on t and on τ_i .

To determine the approximation of order zero $\bar{z}_0(t)$ and $\mathfrak{z}_0^{(i)}(\tau_i)$, $i = \overline{0, p}$, we obtain the systems

(25) $$f(t, \bar{z}_0(t), 0) = 0,$$

(26) $$\bar{z}_0(t) + \mathfrak{z}_0^{(0)}(0) = z_o ,$$

(27) $$\Delta\bar{z}_0(t_i) = I_i(\bar{z}_0(t_i)) - \mathfrak{z}_0^{(i)}(0),$$

(28) $$\dot{\mathfrak{z}}_0^{(i)}(\tau_i) = f(t_i , \bar{z}_0(t_i+0) + \mathfrak{z}_0^{(i)}(\tau_i), 0).$$

From (25) in view of condition H3 we find

$$\bar{z}_0(t) = \varphi(t, \alpha_0(t)),$$

where $\alpha_0(t)$ is an arbitrary for the moment k-dimensional vector-valued function.

Now systems (28), $i = \overline{0, p}$, take the form

(29_i) $$\dot{\mathfrak{z}}_0^{(i)}(\tau_i) = f(t_i , \varphi(t_i , \alpha_0(t_i+0)) + \mathfrak{z}_0^{(i)}(\tau_i), 0),$$

i.e. they are systems of the form (17_i). Let $\Omega^{(i)}(\alpha_0(t_i+0))$ be the respective stable manifolds. In order to satisfy condition (24) we wish to have $\mathfrak{z}_0^{(i)}(0) \in \Omega^{(i)}(\alpha_0(t_i+0))$, $i = \overline{0, p}$.

Set $z_0 = (x_0 , y_0)$, $\varphi = (\varphi_1 , \varphi_2)$. By (26), in order to have $\mathfrak{z}_0^{(0)}(0) \in \Omega^{(0)}(\alpha_0(0))$, we must have

(30) $$y_0 - \varphi_2(0, \alpha_0(0)) = P^{(0)}(x_0 - \varphi_1(0, \alpha_0(0)), \alpha_0(0)).$$

In fact, this is a system of k equations with k unknowns - the components of $\alpha_0(0)$. Assume that

H7. Equation (30) has a solution $\alpha_0(0) = \alpha_0 \in D$.

Thus we determine the initial condition α_0 for the yet unknown function $\alpha(t)$. Now the solution of (29_0) satisfies $\mathfrak{z}_0^{(0)}(\tau_0)\in\Omega^{(0)}(\alpha_0)$ for all $\tau_0 \geq 0$ and the estimate

$$(31)\qquad |\mathfrak{z}_\nu^{(i)}(\tau_i)| \leq c.e^{-\kappa\tau_i}, \quad \tau_i \geq 0,$$

for some positive constants c and κ, and for $i = \nu = 0$.

Analogously, in order to have $\mathfrak{z}_0^{(i)}(0) \in \Omega^{(i)}(\alpha_0(t_i))$, we must have

$$(32_i)\qquad I_{i2}(\varphi(t_i,\alpha_0(t_i))) - \varphi_2(t_i,\alpha_0(t_i+0)) + \varphi_2(t_i,\alpha_0(t_i))$$
$$=P^{(i)}\Big(I_{i1}(\varphi(t_i,\alpha_0(t_i)))-\varphi_1(t_i,\alpha_0(t_i+0))+\varphi_1(t_i,\alpha_0(t_i)),\alpha_0(t_i)\Big).$$

H8. For $i = \overline{1,p}$ from equations (32_i) $\alpha_0(t_i+0)$ can be expressed as functions of $\alpha_0(t_i) \in D$ with values in D:

$$(33)\qquad \alpha_0(t_i+0) = \chi_i(\alpha_0(t_i)).$$

Thus we determine the jump conditions for the unknown function $\alpha_0(t)$. Once $\alpha_0(t)$ is determined, we obtain that the functions $\mathfrak{z}_0^{(i)}(\tau_i)$ also satisfy estimate (31).

The function $\alpha_0(t)$ will be completely determined from the solvability condition for $\bar{z}_1(t)$. The latter function must satisfy the equation

$$\dot{\bar{z}}_0(t) = f_z(t, \bar{z}_0(t), 0)\bar{z}_1(t) + f_\varepsilon(t, \bar{z}_0(t), 0)$$

or

$$(34)\qquad f_z(t, \varphi(t, \alpha_0(t)), 0)\bar{z}_1(t)$$
$$= \varphi_\alpha(t, \alpha_0(t))\dot{\alpha}_0(t) + \varphi_t(t, \alpha_0(t)) - f_\varepsilon(t, \varphi(t, \alpha_0(t)), 0).$$

This is a linear algebraic system of equations whose determinant equals zero. For the solvability of this system it is necessary and sufficient that its right-hand side be orthogonal to the eigenvectors $g_j(t, \alpha_0(t))$, $j = \overline{1, k}$, of the conjugate matrix $f_z^*(t,\varphi(t,\alpha_0(t)),0)$ corresponding to the eigenvalue $\lambda = 0$. Denote by $g(t, \alpha_0(t))$ the $(k \times m)$- matrix with rows the null eigenvectors $g_j(t, \alpha_0(t))$. Then the orthogonality condition can be written down as follows:

$$(35)\qquad \Big(g(t, \alpha_0(t)).\varphi_\alpha(t, \alpha_0(t))\Big)\dot{\alpha}_0(t)$$
$$= g(t, \alpha_0(t))\Big(f_\varepsilon(t, \varphi(t, \alpha_0(t)), 0) - \varphi_t(t, \alpha_0(t))\Big).$$

It is easy to see that the $(k \times k)$ - matrix

$$g(t, \alpha_0(t)).\varphi_\alpha(t, \alpha_0(t))$$

is nonsingular, thus equation (35) can be written in the form

$$(36)\qquad \dot{\alpha}_0(t) = f_0(t, \alpha_0(t)).$$

H9. Equation (36) with initial condition α_0 given by H7 and jump conditions (33) has a solution $\alpha_0(t)$ defined for $0 \le t \le T$ and with values in D.

Now $\bar{z}_0(t) = \varphi(t, \alpha_0(t))$ is completely determined, and so are the boundary functions $\mathfrak{z}_0^{(i)}(\tau_i)$, $i = \overline{0, p}$, provided that $\bar{z}_0(t_i+0) \in \Omega$.

H10. The sets $\{\bar{z}_0(t);\ 0 \le t \le T\}$ and $\{\bar{z}_0(t_i+0)+\mathfrak{z}_0^{(i)}(\tau_i);\ \tau_i \ge 0\}$, $i = \overline{0, p}$, are contained in Ω.

Let T(t) be a matrix such that for $0 \le t \le T$

$$(37) \qquad T^{-1}(t)F(t, \alpha_0(t))T(t) = \begin{pmatrix} \bar{a}_{11}(t) & 0 \\ 0 & 0 \end{pmatrix},$$

where $\bar{a}_{11}(t)$ is a matrix of dimension $(m-k) \times (m-k)$ whose eigenvalues $\lambda_j(t) \equiv \lambda_j(t, \alpha_0(t))$, $j = \overline{1, m-k}$, satisfy (16).

Denote $P_i = \partial_z I_i(\bar{z}_0(t_i))$, $i = \overline{1, p}$,

$$Q_i = T^{-1}(t_i+0)\Big(P_i T(t_i) - \Delta T(t_i)\Big) = \begin{pmatrix} Q_{11,i} & Q_{12,i} \\ Q_{21,i} & Q_{22,i} \end{pmatrix}.$$

H11. The matrices $E+Q_{jj,i}$, $i = \overline{1,p}$, $j = 1,2$, are nonsingular, while $Q_{12,i} = 0$.

This condition implies some kind of compatibility between the impulse effects and the respective stable manifolds.

Now the solution of equation (34) can be written in the form

$$(38) \qquad \bar{z}_1(t) = \bar{\varphi}_\alpha(t)\alpha_1(t) + \tilde{z}_1(t),$$

where $\bar{\varphi}_\alpha(t) \equiv \varphi_\alpha(t, \alpha_0(t))$, $\tilde{z}_1(t)$ is an arbitrary solution of (34), and $\alpha_1(t)$ is a still arbitrary k-dimensional vector-valued function.

For the boundary functions $\mathfrak{z}_1^{(i)}(\tau_i)$, $i = \overline{0, p}$, we obtain equations of the form

$$(39) \quad \dot{\mathfrak{z}}_1^{(i)}(\tau_i) = F_z^{(i)}(\tau_i)\mathfrak{z}_1^{(i)}(\tau_i) + F_z^{(i)}(\tau_i)\bar{\varphi}_\alpha(t_i+0)\alpha_1(t_i+0) + \psi^{(i)}(\tau_i),$$

where $\psi^{(i)}(\tau_i)$, $i = \overline{0, p}$, are known functions satisfying the estimates $|\psi^{(i)}(\tau_i)| \le c.\exp(-\kappa\tau_i)$.

The initial condition for $\mathfrak{z}_1^{(0)}(\tau_0)$ has the form

$$\mathfrak{z}_1^{(0)}(0) = -\bar{z}_1(0) = -\bar{\varphi}_\alpha(0)\alpha_1(0) - \tilde{z}_1(0).$$

The arbitrary up to now vector $\alpha_1(0)$ will be chosen so that to ensure the exponential decay of $\mathfrak{z}_1^{(0)}(\tau_0)$ as $\tau_0 \longrightarrow +\infty$.

By Lemma 3 for $\alpha_1(0)$ we obtain a linear algebraic equation with left-hand side $R(\infty, \alpha_0)\alpha_1(0)$. By virtue of Lemma 5 this equation is uniquely solvable with respect to $\alpha_1(0) = \alpha_1^0$. Then by Lemma 3 and Lemma 4 $\mathfrak{z}_1^{(0)}(\tau_0)$ satisfies the exponential estimate (31) (i=0, ν=1).

Up to now we have found the initial condition α_1^0 for the unknown function $\alpha_1(t)$. The jump conditions for $\alpha_1(t)$ can be found from the equalities

(40) $$\Delta\bar{z}_1(t_i) + \mathfrak{z}_1^{(i)}(0) = \partial_z I_i(\bar{z}_0(t_i))\bar{z}_1(t_i), \quad i = \overline{1, p}.$$

As above, we find that the solutions $\mathfrak{z}_1^{(i)}(\tau_i)$, $i=\overline{1,p}$, of (39) with initial condition found from (40) satisfy the exponential estimate (31) if $\alpha_1(t_i+0)$ is a certain linear function of $\alpha_1(t_i)$:

(41) $$\alpha_1(t_i+0) = A_i\alpha_1(t_i) + B_i \ .$$

Thus we determined the jump conditions for $\alpha_1(t)$. This function will be completely determined from the solvability condition for $\bar{z}_2(t)$. As above we obtain an equation of the form

(42) $$\dot{\alpha}_1(t) = f_1(t, \alpha_1(t)).$$

Here $f_1(t, \alpha_1)$ is a vector-valued function whose components are polynomials of degree two with respect to the components of α_1 .

H12. The impulsive equation (42) with initial condition $\alpha_1(0) = \alpha_1^0$ and jump conditions (41) has a solution $\alpha_1(t)$ for $0 \le t \le T$.

This is the last assumption we make. Now $\bar{z}_1(t)$ and $\mathfrak{z}_1^{(i)}(\tau_i)$, $i=\overline{0,p}$, are completely determined.

The subsequent terms of the asymptotics are determined analogously. On the ν-th step, $\nu \ge 2$, the expression for $\bar{z}_\nu(t)$ will contain an arbitrary function $\alpha_\nu(t)$. First from the conditions $\mathfrak{z}_\nu^{(i)}(\tau_i) \longrightarrow 0$ as $\tau_i \longrightarrow \infty$ we determine $\alpha_\nu(0)$, and $\alpha_\nu(t_i+0)$ as linear functions of $\alpha_\nu(t_i)$, $i = \overline{1, p}$, just as above. Then from the solvability condition for $\bar{z}_{\nu+1}(t)$ we obtain an equation of the form

(43) $$\dot{\alpha}_\nu(t) = f_\nu(t, \alpha_\nu(t)),$$

where for $\nu \ge 2$ the functions $f_\nu(t, \alpha_\nu)$ are linear with respect to α_ν (and different for different ν). The solutions of the linear equations (43) with given initial and linear jump conditions are uniquely determined for $0 \le t \le T$. Thus under the assumptions made the terms of the series (21) $\bar{z}_\nu(t)$, $\mathfrak{z}_\nu^{(i)}(\tau_i)$, $i = \overline{0, p}$, can be determined up to $\nu = n + 1$.

5. JUSTIFICATION OF THE METHOD

Theorem 1. Under conditions H1-H12 there exist positive constants ε_0 and c such that for $\varepsilon \in (0, \varepsilon_0]$ there exists a unique solution $z(t, \varepsilon)$ of problem (12)-(14) on the segment [0, T], which satisfies the inequality

$$(44) \qquad |z(t, \varepsilon) - Z_n(t, \varepsilon)| \le c\varepsilon^{n+1}, \ 0 \le t \le T.$$

Sketch of the proof. In system (12)-(14) we carry out the change of variables

$$(45) \qquad z = w + Z_{n+1}(t, \varepsilon).$$

and obtain the system

$$(46) \qquad \varepsilon\dot{w} = F_z(t, \varepsilon)w + G(t, \varepsilon, w),$$

$$(47) \qquad \Delta w(t_i) = P_i(\varepsilon)w(t_i) + p_i(\varepsilon, w(t_i))$$

with initial condition

$$(48) \qquad w(0, \varepsilon) = 0,$$

where

$$F_z(t, \varepsilon) = f_z(t, Z_1(t, \varepsilon), \varepsilon),$$

$$G(t, \varepsilon, w) = f(t, w + Z_{n+1}(t,\varepsilon), \varepsilon) - F_z(t,\varepsilon) - \varepsilon \frac{dZ_{n+1}(t,\varepsilon)}{dt},$$

$$P_i(\varepsilon) = \partial_z I_i(\bar{z}_0(t_i) + \varepsilon\bar{z}_1(t_i)), \ i = \overline{1, p},$$

$$p_i(\varepsilon, w(t_i)) = I_i\left(w(t_i) + Z_{n+1}^{(i-1)}(t_i, \varepsilon)\right)$$

$$+ Z_{n+1}^{(i-1)}(t_i, \varepsilon) - Z_{n+1}^{(i)}(t_i, \varepsilon) - P_i(\varepsilon)w(t_i).$$

The function $G(t, \varepsilon, w)$ enjoys the following two properties:

1. $G(t, \varepsilon, 0) = O(\varepsilon^{n+2})$.

2. For any two positive constants c_1 and ε_1 (ε_1 small enough) there exist constants $c_0 > 0$, $\varepsilon_0 \in (0, \varepsilon_1]$ such that for any w_1 , w_2 satisfying for $0 \le t \le T$, $0 < \varepsilon \le \varepsilon_1$ the inequalities

$$|w_j(t, \varepsilon)| \le c_1\varepsilon^2, \ j = 1, 2,$$

for $0 \le t \le T$, $0 < \varepsilon \le \varepsilon_0$ we have

$$|G(t, \varepsilon, w_1(t,\varepsilon)) - G(t, \varepsilon, w_2(t,\varepsilon))|$$

$$\le c_0\varepsilon^2 \max_{0\le t\le T} |w_1(t,\varepsilon) - w_2(t,\varepsilon)|.$$

For the functions p_i , $i = \overline{1, p}$, the analogous properties will be denoted by $1'$, $2'$. We omit their formulation. For the verification of properties 1, 2 and $1'$, $2'$ see [1] (the noncritical case) and [2] (the critical case without impulses).

In (46)-(48) we carry out a new change of the variables

$$w = T(t)\begin{pmatrix} u \\ v \end{pmatrix}, \tag{49}$$

where $u \in \mathbb{R}^{m-k}$, $v \in \mathbb{R}^{k}$, the matrix $T(t)$ satisfies (37). Then we replace the impulsive system obtained by an equivalent system of integral equations which we solve by the method of successive approximations starting from $u_0 = 0$, $v_0 = 0$.

We can choose ε_1 so small that for $0 < \varepsilon \le \varepsilon_1$ the convergence of the method of successive approximations is ensured. Moreover, the solution of the system of integral equations satisfies estimates of the form

$$|u(t, \varepsilon)| \le c\varepsilon^{n+1}, \quad |v(t, \varepsilon)| \le c\varepsilon^{n+1}.$$

By the changes of variables (45) and (49) we obtain a solution $z(t, \varepsilon) = (x(t, \varepsilon), y(t, \varepsilon))$ of (12)-(14) satisfying

$$|z(t, \varepsilon) - Z_{n+1}(t, \varepsilon)| \le c\varepsilon^{n+1}, \quad 0 \le t \le T, \ 0 < \varepsilon \le \varepsilon_0$$

with another positive constant still denoted by c. Since

$$|Z_n(t, \varepsilon) - Z_{n+1}(t, \varepsilon)| \le c\varepsilon^{n+1},$$

the solution $z(t, \varepsilon)$ satisfies (44).□

ACKNOWLEDGEMENT

The present investigation is supported by the Bulgarian Ministry of Education and Science under Grants MM-7 and MM-8.

References

1. S.G. Hristova and D.D. Bainov, Riv. Matem. Pura Appl., 10, 67-87 (1991).

2. A.B.Vasil'eva and V.F.Butuzov, Singularly Perturbed Equations in the Critical Case, Izd.Moscow Univ., Moscow (1978) (Russian).

Third International Colloquium on Differential Equations pp. 45-54 (1993)
D. Bainov and V. Covachev (Eds)

RECENT DEVELOPMENTS ON NUMERICAL INTEGRATORS FOR CURVES ON PRESCRIBED SURFACES

José Manuel Ferrándiz and M. Teresa Pérez
E.T.S. de Ingenieros Industriales. Paseo del Cauce s/n.
E-47011 Valladolid, SPAIN

Abstract
The first part of this paper is a survey of both analytic and numerical methods that have been developed mainly in Celestial Mechanics to take advantage of first integrals that a differential system might posses, giving rise to different stabilization methods or new numerical codes.

Guided by the same idea, in the second part we introduce a new procedure to improve the numerical results of difference schemes. It is specially efficient for IVP with solutions on perturbed spheres, as it is shown through some numerical experiments.

Keywords: Numerical Integration, First Integrals.

1. INTRODUCTION

From the mid-60's to the beginning of the 80's several authors suggested stabilization procedures for improving the accuracy of numerical integration methods for ordinary differential equations. The main idea was to make use, in one way or another, of the first integrals that the problem in question might have. When a system of differential equations admits a first integral, its value can be used to estimate the reliability of the numerical integration.

It is clear that if the value of the integral is not reasonably well preserved, we must conclude that the integration is not sufficiently accurate. On the other hand, an improvement in the approximation to the integral is usually interpreted as an indication of improvement in the numerical solution, although, of course the errors in the integral and in the solution must not be confused, since the former can be small but the latter enlarged due to tangential displacements —for example, simple changes of phase.

That the satisfaction of the first integral is a necessary but insufficient condition for guaranteeing the veracity of the integration should not lead us to think that its achievement is a waste of effort. In fact, a large number of systems with practical interest are unstable in the Lyapunov sense; in general, all the gravitational systems of interest in Celestial Mechanics share this property. Due to this, if the error arising from the deviation with respect to the integrals remains in the solution during a process of numerical integration, the solutions will diverge faster than would the solution without such a source of error, and with time it would often be impossible to maintain the accuracy at a reasonable level. More detailed discussions on this theme can be found in Miller [1] and in Szebehely [2].

Guided by these ideas, Nacozy ([3], [4]) developed a technique for controlling the error in the integrals and for achieving more efficient numerical solutions. In this technique any method whatsoever can be applied to the numerical integration of the equations of a system with p first integrals. After each step, the computed solution will not be on surfaces, but at a certain distance, generally small, from them. In order to return the point to the integrals a least–squares correction of the values obtained is calculated, which supposes the solution's return to the relevant hypersurface along the normal direction with a second order error with respect to the resulting error before correction. When there are several integrals, as happens with the problem of various bodies, where $p = 10$, the correction needs the insertion of a matrix $p \times p$. Even so, in the experiments described by Nacozy, increased accuracy was achieved maintaining the calculation time. Other corrections to constrain the computed solutions of gravitational n–body systems to remain on the integrals have previously been performed by Aarseth [5].

Nacozy's procedure does not require a change in the equations to integrate nor in the numerical scheme, only the use of a special corrector procedure depending on the problem. Nevertheless, the majority of the methods thought up to conserve the integrals consists in replacing the equations of motion with others that have better properties towards this end.

Possibly one of the simplest and most successful examples is in the stabilization of equations of motion corresponding to two–body problems in KS variables. A very complete description of the "KS methods" (as they are called in Celestial Mechanics), developed from the Kustaanheimo transformation, can be seen in the book by Stiefel and Scheifele [6]. For our purposes, it is enough to mention that the substitution of the energy integral in the equations of motion written in KS coordinates allows us to transform the equations into a perturbed 4–dimensional harmonic oscillator, with constant frequency equal to 1/2, which produces a notable stabilizing effect. The price we have to pay is the addition of an equation for the slow variation of the energy, which should be numerically integrated. In this way the original system has been embedded in a wider one better adapted to numerical integration, whose solutions allow us to recuperate those of the former.

The same procedure was used by Baumgarte [7] to achieve the stabilization –in the same sense as before– of the Keplerian motion in Cartesian variables. Baumgarte [8] later, showed how the first integral could be satisfied with greater accuracy by adding an appropriate control term to the equation via the process he called *dissipative stabilization.* If the first integral is taken to be a surface, this term can be considered as a fictitious force perpendicular to it, whose magnitude increases with the distance to the surface. In fact, if $h = h_0$ is the said integral he takes $f = \dot{h} + \gamma(h - h_0) = 0$ with $\gamma > 0$ and constant, as a Gaussian constraint for the equation. In this way as

$$\frac{d}{dt}(h - h_0)^2 \leq 0$$

the error in the first integral decreases with the time and therefore also as the integration advances.

The above–mentioned author introduced various other modifications of the equations of motion, for example, through the manipulation of the Langrangian or of the Hamiltonian ([9], [10]).

On these lines, the most interesting method from a conceptual point of view seems to be the so–called dissipative or gradient stabilization (Sigrist [11], Baumgarte [7], Baumgarte and Stiefel [9]), which was later studied theoretically by Kirchgraber ([12],

[13]) using tools reminiscent of the theory of invariant manifold. More information on this kind of techniques and their application to gravitational problems can be seen by Aarseth(1988).

A notably different approach to the problem was made by Zubov [15], who instead of modifying the differential equations as the above-mentioned authors had done, modified the numerical integration method. Such a modification is carried out by making use of a matrix formed from the gradients of the first integral that the system possesses (as in Nacozy), in such a way that the resulting equation in differences has precisely, as first integrals, the result of discretisizing the first integrals of the problem.

Those multistep methods known as circulary exact algorithms can also be seen as special cases of numerical schemes that preserve first integrals, although they were introduced by Lambert and McLeod [16]in a different context. In their conception, the emphasis is put on the geometry of the solution curve support and, infact, the authors talk about trajectory problems, such as those where the interest lies in the curve traced by the solution instead of the actual correspondence between the values of the independent variable and the points of that curve. The first scheme, of order two, permitted the exact integration of circumferences on the plane since the points generated by the algorithm are on the same circumference as the starting points. Other versions are able to integrate conics or circumferences in R^n ([16]), always with a low order. The study of the error and, in general, of the properties of the methods are carried out from a mainly geometric point of view, based on the elements of the solution curve ([18], [19]).

Inspired by the above, the authors recently developed spherically exact methods ([20], [21], [22]) with the property that the points generated by the algorithm remain on the spherical surface provided the starting values lie on it, naturally discounting round-off errors. The reason for their design is that in certain problems it is interesting to take advantage of the fact that the solution curves are spherical. Such is the case with orbital problems formulated in focal variables (Burdet [23], Ferrándiz [24]), in which part of the coordinates correspond to the direction vector of the particle.

The aforementioned methods were also constructed in a geometric way, from standard multistep codes, such as the Adams ones, by means of a projection process with respect to the normal and tangent manifolds to the invariant surface, which is a sphere in the Euclidean space of arbitrary dimensions. With this procedure higher orders than for those of the usual circularly exact ones are obtained and the geodesics are integrated in an exact way; for any spherical curve, the smaller its geodesic curvature is the better the integration, since the local truncation error contains it as a factor. The procedure used in the construction of these algorithms can be generalized to other surfaces although a great deal of the simplicity of formulation and putting into practice are lost, as a consequence of the high symmetry of the sphere. In fact there are implicit R–K methods that preserve arbitray surfaces for certain problems as Cooper [25] showed and that were called orbitally stable by this author.

Despite the difficulties that we have pointed out, in certain cases the invariant manifolds are not real Euclidean spheres but can be seen as slight deformations of them as they arise from perturbation problems. This is the case in two–body problems in the KS approach, mentioned before. This gave us the idea to look for methods that, although they do not adapt exactly to the first integral, they do allow us to take advantage of its proximity to a sphere in order to suppress from the error the terms that come from the curvature of this surface and which are easily located without any additional calculations. In this way we can get the first terms of the truncation error to contain a factor which order is that of the small parameter of the perturbation and that experience has shown us to be normally fairly beneficial.

In the rest of this paper we give a brief description of the spherically exact methods and of the ones last mentioned, to be completed with a few numerical examples in section 4.

2. SPHERICALLY EXACT METHODS

Although in [20] and [21] some particular methods of this kind appear, we are going to pay attention only to those described in [22]. The leading idea of this construction is very simple: for a curve, $\mathbf{y}(.)$, lying on a sphere, parametrized by its arc length, if s_0 and s are two values of the parameter for which the Euclidean distance between the corresponding points is $\|\mathbf{y}(s_0) - \mathbf{y}(s)\| = h$, the component of $\mathbf{y}(s)$ in the normal direction to the surface considered in $\mathbf{y}(s_0)$ is exactly $(1 - \frac{h^2}{2})\mathbf{y}(s_0)$. Then it is possible to remove the error in this direction from any approximation to $\mathbf{y}(s)$. Besides, it is possible to force the approximation to lie on the surface in the following way: if $\mathbf{u}$ denotes a previous approximation to $\mathbf{y}(s)$, then we can generate a new one by means of the formula

$$(1 - \frac{h^2}{2})\mathbf{y}(s_0) + h\sqrt{1 - -\frac{h^2}{4}}\frac{\mathbf{c}}{\|\mathbf{c}\|}, \tag{1}$$

where $\mathbf{c} = \mathbf{u} - < \mathbf{u}, \mathbf{y}(s_0) > \mathbf{y}(s_0)$ and $< .,. >$ stands for the dot product. The new point is at Euclidean distance h from $\mathbf{y}(s)$ and we can guarantee that the error in the tangential direction to $\mathbf{y}(s_0)$ is small relative to the transversal error.

Based on these ideas, we have designed a family of multistep methods to integrate normalized differential equations in R^n

$$\mathbf{y}' = \mathbf{f}(\mathbf{y}), \quad \|\mathbf{f}\| = 1,$$

in the case that the solutions are spherical curves. The algorithm can be written as

$$\mathbf{y}_{j+m} = \mathbf{y}_{j+m-1} + hF(\mathbf{y}_j, \ldots, \mathbf{y}_{j+m}), \tag{2}$$

where the function F has an expression of the form

$$F(\mathbf{y}_j, \ldots, \mathbf{y}_{j+m}) = -\frac{h}{2}\mathbf{y}_{j+m-1} + \sqrt{1 - -\frac{h^2}{4}}\frac{\mathbf{c}_{j+m-1}}{\|\mathbf{c}_{j+m-1}\|},$$

$\mathbf{c}_{j+m-1}$ being the component in the hyperplane tangent to $\mathbf{y}_{j+m-1}$ of the vector that plays the role of the previous approximation quoted above. Usually we select the coefficients in such a way that the substitution of F by u in (2), with

$$u = \sum_{i=0}^{m} \alpha_i[\mathbf{f}(\mathbf{y}_{j+i}) - \mathbf{f}(\mathbf{y}_{j+i-1})] + \mathbf{f}(\mathbf{y}_{j+m-1}),$$

would produce a standard code, such as the Adams type ones. Thus we can envisage the algorithm (2) as the result of a certain projection of the former one along the normal and tangent manifolds in a point of the sphere.

These schemes leave the surface invariant, a property that we have called spherical exactness. Besides two consecutive points are always at Euclidean distance h. Their numerical behaviour is also good: they are consistent, their stability can be proved using the main theorem of Grigorieff [26], and thus they converge. The principal term of the local truncation error has a component neither in the tangential direction nor in the normal one to the spherical surface to $\mathbf{y}(s_{j+m-1})$, provided that the values of the parameter are chosen to make the points equispaced —the so-called chord criterium.

Moreover, the methods are exact along geodesics, and the local error is factorized by the geodesic curvature and its derivatives. So the codes are suitable for integrating orbits near geodesics, and they allow us to gain accuracy with respect to the usual algorithms.

3 A MODIFICATION OF MULTISTEP METHODS

Despite the good behaviour that those methods described in the former section show, they have some limitations. The most important one comes from the fact that they can only be applied with success to a rather reduced kind of problems: those which have solutions on a sphere of a Euclidean space of arbitrary dimension. Secondly, they are not advisable when the exact correspondence between an independent varible, fixed forehand, and the calculated points of the solution are to be known. In this section we present a different alternative to modify the classic multistep methods, that turned out to be useful not only in the same problems as the spherically exact algorithms, but also to integrate curves on surfaces near spheres, that is to say, surfaces obtained from spheres by means of small deformations. In order to simplify the algorithms, it is enough to parametrize the solutions with their arc length, then the computed points of the solution correspond to values of this variable. As we will see later in the section dedicated to numerical examples, in many cases these algorithms produced an improvement in accuracy similar to the spherically exact ones —naturally, with no additional noticeable computational cost– a reason that makes them a good choice.

In a first approach, we introduce these new methods in an intuitive way. For this purpose we start by considering an initial IVP in R^3

$$\mathbf{x}' = \mathbf{f}(t,\mathbf{x}), \quad t \in [a,b],$$

$$\mathbf{x}(a) = \mathbf{w},$$

where $\mathbf{f}$ is a field tangent to the sphere S^2, and normalized in such a way that $\|\mathbf{f}(t,\mathbf{x})\| = 1$, $\forall t \in [a,b]$, the solution of this problem being a curve on S^2 parametrized by its arc length. Let us denote by $\{\mathbf{t},\mathbf{N},\mathbf{T}\}$ the *Darboux-Ribaucour* trihedrom in any point of the solution curve, where $\mathbf{t} = \mathbf{f}(t,\mathbf{x})$ is the tangent vector, $\mathbf{N}$ the inwards normal in $\mathbf{x}$ and $\mathbf{T}$ completes the orthonormal frame. Let us recall the following equalities:

$$\begin{pmatrix} \mathbf{t} \\ \mathbf{N} \\ \mathbf{T} \end{pmatrix}' = \begin{pmatrix} 0 & k_n & k_g \\ -k_n & 0 & \tau_g \\ -k_g & -\tau_g & 0 \end{pmatrix} \begin{pmatrix} \mathbf{t} \\ \mathbf{N} \\ \mathbf{T} \end{pmatrix},$$

where $k_n = k_n(t)$ is the normal curvature of the surface, $k_g = k_g(t)$ the geodesic curvature and $\tau_g = \tau_g(t)$ the geodesic torsion. For a curve on S^2 we have $k_n(t) = 1$ and $\tau_g(t) = 0$, $\forall t \in [a,b]$, and the former formulae reduce to

$$\begin{aligned} \mathbf{t}' &= \mathbf{N} + k_g \mathbf{T}, \\ \mathbf{N}' &= -\mathbf{t}, \\ \mathbf{T}' &= -k_g \mathbf{t}. \end{aligned}$$

If the curve is a geodesic then $k_g = 0$ and the successive derivatives of $\mathbf{x}$ can be easily calculated. This can be done in terms of $\mathbf{f}$ and $\mathbf{x}$ itself because the outwards normal field to the sphere is given by $\mathbf{x}$. More precisely, we have

$$\mathbf{x}^{(2k)}(t) = (-1)^k \mathbf{x}(t) \text{ and } \mathbf{x}^{(2k)}(t) = (-1)^k \mathbf{f}(t, \mathbf{x}(t)). \quad (3)$$

The knowledge of the derivatives of the function allows us to modify the numerical methods with the aim of reducing the error, as we described in what follows.

Given a linear multistep method (ρ, σ), represented by the difference equation

$$F_n(X_n, \ldots, X_{n+k}; h) = \alpha_k X_{n+k} + \ldots + \alpha_0 X_n$$

$$-h\{\beta_k f(t_{n+k}, X_{n+k}) + \ldots + \beta_0 f(t_n, X_n)\}$$

with order of consistency p and whose local truncation error can be written as follows

$$F_n(\mathbf{x}(t_n), \ldots, \mathbf{x}(t_{n+k}); h) = C_{p+1} h^{p+1} \mathbf{x}^{(p+1)}(t_{n+k-1})$$

$$+C_{p+2} h^{p+2} x^{(p+2)}(t_{n+k-1}) + \ldots + C_{p+l} h^{p+l} x^{(p+l)}(t_{n+k-1}) + O(h^{p+l+1}),$$

we propose a new class of methods:

$$F'_n(X_n, \ldots, X_{n+k}; h) = \alpha_k X_{n+k} + \ldots + \alpha_0 X_n - h\{\beta_k \mathbf{f}(t_{n+k}, X_{n+k})$$

$$+\beta_{k-1}(\mathbf{f}(t_{n+k-1}, X_{n+k-1}) + \mathbf{g}(t_{n+k-1}, X_{n+k-1})) + \beta_{k-2} \mathbf{f}(t_{n+k-2}, X_{n+k-2} + \ldots + \beta_0 \mathbf{f}(t_n, X_n)\}$$

where $\mathbf{g}$ is a function of the form

$$\mathbf{g}(t, \mathbf{x}) = h^{p+1}\{C_{p+1} \mathbf{g}^1(t, \mathbf{x}) + h C_{p+2} \mathbf{g}^2(t, \mathbf{x}) + \ldots + h^{l-1} C_{p+l} \mathbf{g}^l(t, \mathbf{x})\},$$

C_i being precisely the constants for the local truncation error of (ρ, σ). Then the local truncation error for the new methods can be written as

$$F'_n(\mathbf{x}(t_n), \ldots, \mathbf{x}(t_{n+k}); h) = C_{p+1} h^{p+1} [\mathbf{x}^{(p+1)}(t_{n+k-1}) - \mathbf{g}^1(t_{n+k-1}, \mathbf{x}(t_{n+k-1}))] \quad (4)$$

$$+ \ldots + C_{p+l} h^{p+l} [\mathbf{x}^{(p+l)}(t_{n+k-1}) - \mathbf{g}^l(t_{n+k-1}, \mathbf{x}(t_{n+k-1}))] + O(h^{p+l+1}).$$

It is now evident that if the functions $\mathbf{g}^1$ to $\mathbf{g}^l$ are selected to be the corresponding derivatives of the geodesic, what by (3) can be done without any further calculations, then an increase in the accuracy of order l can be achieved in this case.

The interest of the construction above lies in the fact that it can be applied to the case of systems having first integrals near spheres, i.e., to any perturbed oscillator in $2n$ dimensions with the sole requirement that the trajectories are reasonably near to the solutions of the unperturbed oscilator. Equations (3) are no longer satisfied but when the corresponding derivatives are approximated by means of these equalities the error made can be considered to be of the order of a small parameter ϵ responsible of the perturbation. In this way in the righthand side of (4), the local error admits an asymptotic expansion of the form

$$\epsilon\{O(h^{p+1}) + \ldots + O(h^{p+l})\} + O(h^{p+l+1})$$

and the accuracy of the method can be increased the order of ϵ with a suitable choice of the number of corrector terms. In practice this can be seen as an increase of order. Naturally, the procedure we have just described does not damage the good properties of consistency, stability and convergence of the algorithm modified.

Some results concerning the asymptotic behaviour of the error analogous to those of Henrici [27] can be proved.

4 NUMERICAL EXPERIMENTATION

Several numerical experiments have been carried out to test the behaviour of the schemes we deal with in this paper. In this section some of the results obtained are shown. We have selected three different systems of equations that we describe next.

Problem 1.

$$\begin{aligned} x_1' &= \frac{x_1 x_3}{\sqrt{x_1^2 + x_2^2}}, \\ x_2' &= \frac{x_2 x_3}{\sqrt{x_1^2 + x_2^2}}, \\ x_3' &= \sqrt{x_1^2 + x_2^2}. \end{aligned}$$

The solutions of this system are geodesics on S^2, that is, meridians.

Problem 2.

$$\begin{aligned} x_1' &= -x_2 - \frac{\omega x_1 x_3}{x_1^2 + x_2^2}, \\ x_2' &= x_1 - \frac{\omega x_2 x_3}{x_1^2 + x_2^2}, \\ x_3' &= \omega. \end{aligned}$$

Its solutions are loxodroms on S^2. The parameter ω gives the angle of inclination, and thus the geodesic curvature, with respect to the meridians.

Problem 3 (Henon–Heiles)

$$\begin{aligned} p_1' &= q_1 + 3\alpha q_1 - \beta q_2^2, \\ q_1' &= p_1, \\ p_2' &= q_2 - 2\beta q_1 q_2, \\ q_2' &= p_2. \end{aligned}$$

These equations are derived from the Hamiltonian

$$H = \frac{1}{2}(p_1^2 + q_1^2) + (p_2^2 + q_2^2) + \alpha q_1^3 - \beta q_1 q_2^2,$$

α and β being perturbation parameters. The solutions for this problem lie on perturbed spheres if α and β do not vanish.

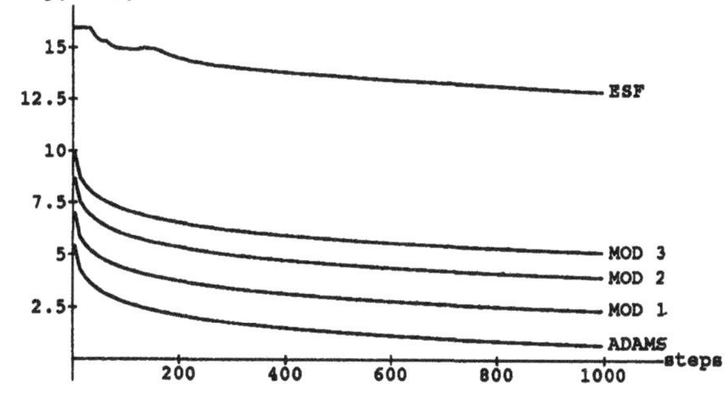

Figure 1: prob.1, order 3, h=0.1

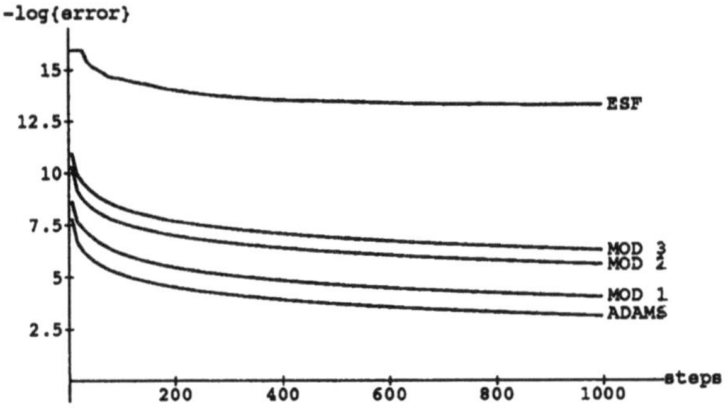

Figure 2: prob.1, order 5, h=0.1

Among the multistep methods we have selected the Adams–Bashforth–Moulton algorithms for their good properties of consistency and stability and their popularity. These have been modified and implemented in PECE mode, also they have been used as reference in the comparation. Double precision has been used in all the experiments.

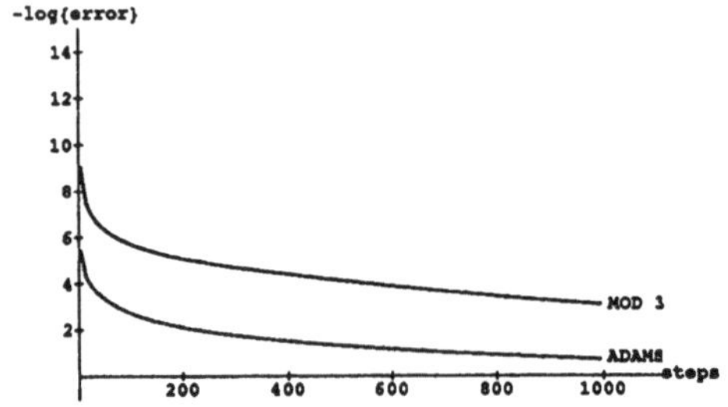

Figure 3: prob. 2, ω=0.001, order 3, h=0.1.

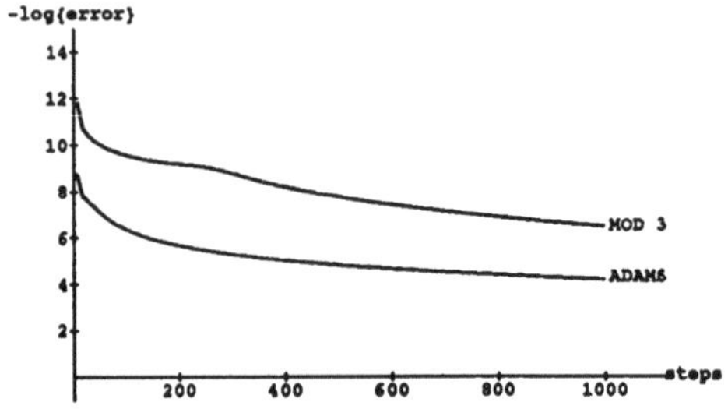

Figure 4: prob. 2, ω=0.001 order 6, h=0.1.

Figures 1 and 2 show the results of the integration of problem 1 starting from the point $(1,0,0)$. The horizontal axis displays the number of steps, and the vertical one the opossite of the decimal logaritm of the error, calculated with respect to a sufficiently accurate numerical solution computed by a RKF method with error control.

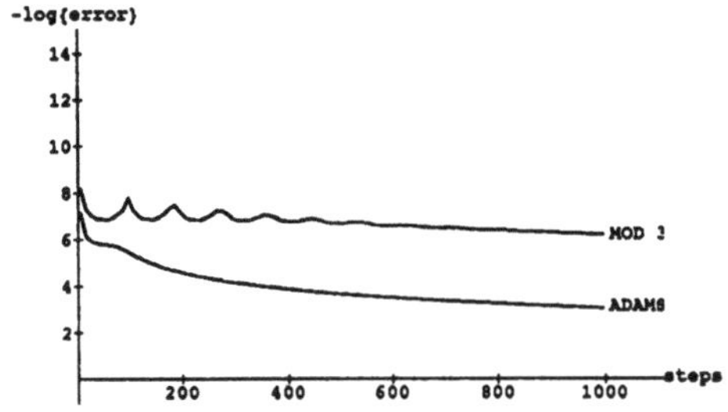

Figure 5: prob. 3, $\alpha = \beta$=0.01, order 4, h=0.1.

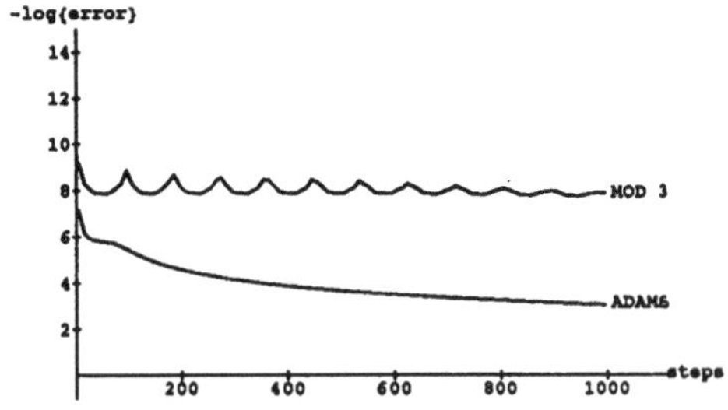

Figure 6: prob. 3, $\alpha = \beta$=0.001, order 4, h=0.1.

The spherically exact method (ESF) is exact except for round-off errors. The new schemes MOD 1, MOD 2 and MOD 3 —where one, two and three terms of the error have been modified, respectively— behave as it can be expected from what we said in section 3, improving the results obtained with the Adams (ADAMS) code.

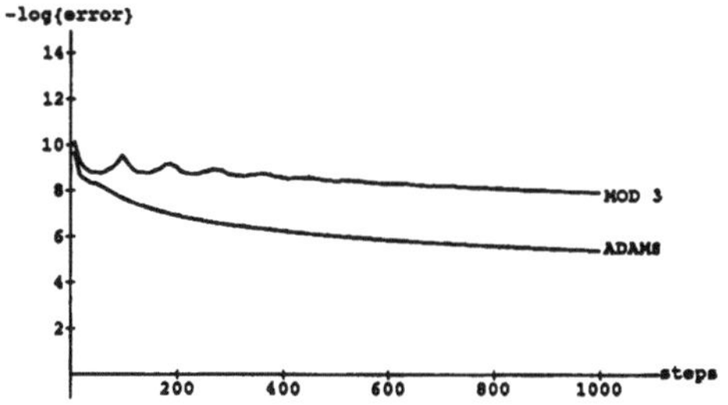

Figure 7: prob. 3, α = β=0.01, order 6, h=0.1.

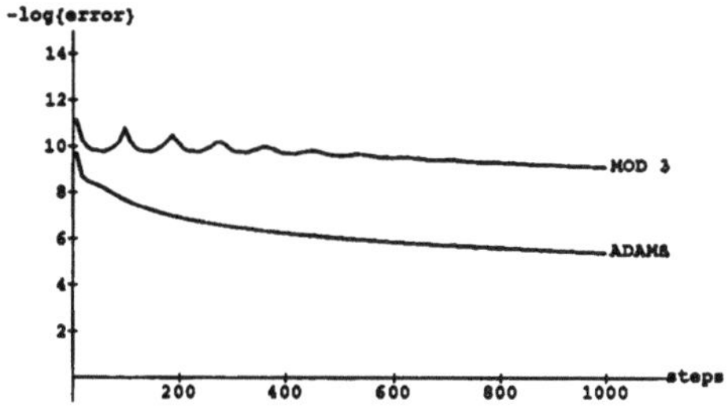

Figure 8: prob. 3, α = β=0.001, order 6, h=0.1.

Figures 3 and 4 correspond to problem 2 with initial value $(1,0,0)$. In this case the results given by the spherical code are almost indistinguishable from those obtained by MOD 3, for that reason we have not plotted them. Figures 5 to 8 show results for problem 3 with initial condition $(1,0,1,0)$. It can be seen that a change in the perturbation affect the results of the new methods: the smaller it is the better the integration is performed, while the Adams give the same results, of course. This behaviour was expected from the way the methods have been derived.

Acknowledgement

The numerical methods by the authors were developed with finantial support from the CICYT (Spain) under Project EP.88-0541.

REFERENCES

[1] R.H. Miller, *Astrophys. J.*, **140**, 250 (1964).

[2] V. Szebehely, Bull. Astron., **3**, 33 (1968).

[3] P. Nacozy, *Astrophys. Space Sci.*, **14**, 40 (1971).

[4] P. Nacozy, in: *Gravitational N-Body Problem*, M.Lecar (ed.), Reidel, p.153 (1972).

[5] S.J. Aarseth, *Monthly Notices Roy. AStron. Soc.*, **132**, 35 (1966).

[6] E. L. Stiefel and G. Scheifele, *Linear and Regular Celestial Mechanics*, Springer, New York (1971).

[7] J. Baumgarte, *Celest. Mech.*, **5**, 490 (1972).

[8] J. Baumgarte, *Celest. Mech.*, **8**, 223 (1973).

[9] J. Baumgarte and E. Stiefel, *Celest. Mech.*, **10**, 71 (1974).

[10] J. Baumgarte, *Celest. Mech.*, **13**, 247 (1976).

[11] N. Sigrist, *ZAMP*, **25**, 37 (1974).

[12] U. Kirchgraber, *ZAMP*, **30**, 272 (1979).

[13] U. Kirchgraber, *Celest. Mech.*, **21**, 225 (1980).

[14] S.J. Aarseth, in:*The Few Body Problem*, M.J: Valtonen (ed.), pp. 287-307 (1988).

[15] V.I. Zubov, *Diff. Eqn.*, **11**, 1685 (1976).

[16] J.D. Lambert and R.J.Y. McLeod, in: *Numerical Analysis Proceedings* (Dundee). Springer, New York (1980), pp. 83-87.

[17] D.P. Laurie, in:*Proceedings of Sixth South African Symposium in Numerical Analysis*, G.R. Joubert (ed.) Durban: Computer Science Department Univ. Natal (1980), pp. 39-45.

[18] R.J.Y. McLeod and J.M. Sanz–Serna, *IMA J. of Numer. Anal.*, **2**, 357 (1982).

[19] J.M. Sanz–Serna, *Numer. Math.*, **45**, 173 (1984).

[20] J.M. Ferrándiz and M.T. Pérez, in: *Actas del X C.E.D.Y.A.* Univ. Valencia (Spain) (1990), pp. 105-109.

[21] J.M. Ferrándiz and M.T. Pérez, in: *Predictability, Stability and Chaos in N-Body Dynamical Systems*, A.E. Roy (ed.) Plenum Publishing Coorporation, NATO ASI Series C (1990), pp. 523-532.

[22] J.M. Ferrándiz and M.T. Pérez, *A Family of Multistep Methods to Integrate Orbits on Spheres* (preprint).

[23] C.A. Burdet, *J. reine und angew. Math.*, **238**, 71 (1969).

[24] J.M. Ferrándiz, *Celest. Mech.*, **41**, 343 (1988)

[25] G.J. Cooper, *IMA J. of Numer. Anal.*, **7**, 1 (1987).

[26] Grigorieff *Numer. Math.*, , (198)

[27] P. Henrici, *Error Propagation for Difference Methods*, Robert E. Krieger Publishing Company, New York (1977), pp.11-22.

Third International Colloquium on Differential Equations pp. 55-62 (1993)
D. Bainov and V. Covachev (Eds)

ON THE CONTACT ANGLE IN CAPILLARITY

ROBERT FINN
Mathematics Department
Stanford University
California 94305 USA

Abstract: This paper offers a brief outline of macroscopic theory implying and implied by the condition of constant contact angle for capillary surfaces. New procedures are described that are designed to use a discontinuous (or nearly discontinuous) dependence on data as a way to test experimentally for the significance of the condition as an intrinsic property of materials, and to obtain precise measurement of the angle in the event of an affirmative result. The discussion is based on a mathematically exact treatment of the classical governing equations; no approximations are employed.

Keywords: Capillarity, contact angle, canonical proboscis

The modern macroscopic theory of capillary surfaces is based on ideas introduced by Young in 1805, by Laplace in 1806, and by Gauss in 1830. These authors studied the physical configuration adopted by the free (capillary) surface S separating two adjacent immiscible fluids (or fluid and gas) that are supported in part by a rigid boundary Z in the presence (or absence) of a gravity field g, see Figure 1. Young [1] gave a heuristic reasoning to support the view that S meets Z in an angle γ that depends only on the materials, and in no way on the geometry of the configuration or on the gravity field.

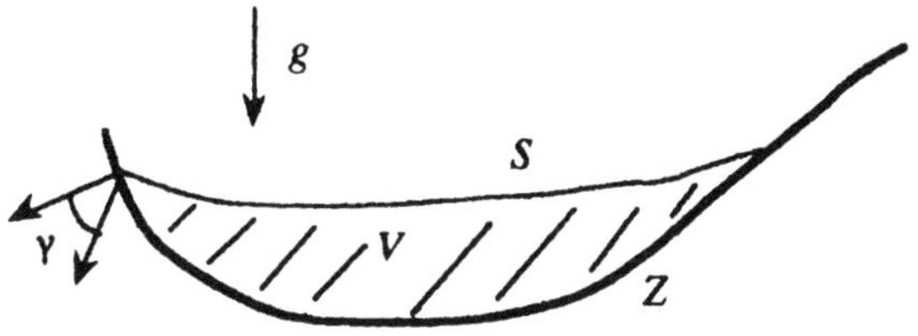

Figure 1: Capillary free surface S; support surface Z

Laplace [2] derived a partial differential equation satisfied locally on the interior of S, and Gauss [3] derived both these results as formal consequences of Johann Bernoulli's Principle of Virtual Work. Specifically, the first variation of the quantity (virtual energy)

$$E \equiv \sigma\{|S| - \beta|S^*| + \kappa\int z\,dV + \lambda V\} \tag{1}$$

must vanish among all variations that preserve the geometrical structure; here σ is the "surface tension", β the "relative adhesion coefficient", S^* the "wetted" part of Z, κ the "capillarity constant" (proportional to g), V the fluid volume, and λ a Lagrange parameter corresponding to the volume constraint. For a formal derivation and more precise definitions, see e.g. [4], Chapter 1. We are led to the equation

$$\Delta\vec{x} = (\kappa u + \lambda)\vec{N} \tag{2}$$

on S, and boundary condition

$$\cos\gamma = \beta \tag{3}$$

on Z. The left side of (2) is the Laplace-Beltrami operator (intrinsic Laplacian) on S; $\vec{x}$ is the position vector and u the height on S, and $\vec{N}$ a unit normal.

A particular case of central classical interest is that of the "capillary tube", in which Z is a vertical cylinder and S can be expressed in the non-parametric form $z = u(x, y)$. See Figure 2. Denoting by Ω a horizontal section of Z, we then obtain the equation

$$\operatorname{div} Tu = \kappa u + \lambda; \quad Tu = \frac{Du}{\sqrt{1 + |Du|^2}} \tag{4}$$

in Ω, with

$$\nu \cdot Tu = \cos\gamma \tag{5}$$

on $\Sigma = \partial\Omega$, ν being unit exterior normal on Σ.

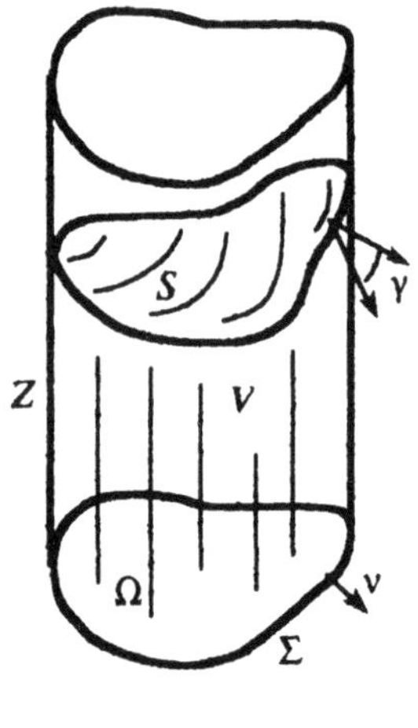

Figure 2: Free surface S in capillary tube

The equation (4) is precisely the one originally given by Laplace. It follows indirectly from observations of Young that the divergence expression in (4) is twice the mean curvature H of S; thus (4) achieves an independent geometrical interest, apart from its physical motivation. The correctness of (4) as a description of reality has to our knowledge never been challenged, although its scientific utility was ridiculed by Young [5], in the context of his general disdain for the mathematical method (and more particularly for those who practiced it).

In fact, to our knowledge the only attempt that has been made to examine the applicability of (4) experimentally appears in the paper of Bashforth and Adams [6]; that investigation was limited to a relatively simple rotationally symmetric configuration and combines the test of the theory with that of a new numerical method. In the context considered, the test did appear to yield a definitively positive result, on both counts.

The boundary condition (5) has also a clear geometrical interest; however its correctness as a description of reality has been - - and still remains - - a matter of continuing controversy, see e.g. the survey by Dussan V [7]. (For a derivation of (5) based on molecular considerations see, e.g. [8] and the references cited there.) The controversy can be traced directly to the difficulty in obtaining repeatable measurements, and different investigators have obtained notably disparate values for presumably identical materials.

Certainly, some of the difficulty is attributable to "hysteresis" forces, arising from resistance to motion of the fluid along a solid surface. This view of the matter has led to introduction of the notions of "advancing" and "receding" contact angles, and these in turn are assumed to depend on the speeds of relative motion (and perhaps also on other factors). From this physical point of view the theory becomes exceedingly complicated, and clear explicit workable principles have yet to emerge. In the limited context of equilibrium configurations, a possible unifying principle has been proposed in [9], [10]; apparently, the matter has not been pursued further along such lines.

The work described in the present article represents an attempt to address the original question, as to the existence of an intrinsic physical angle, by introducing situations in which the adhesion forces that are the basis for the classical theory are overwhelmingly large in respect to the resistance forces. We consider the "capillary tube" configuration described above, in a gravity-free environment (e.g., in space flight). Then (4) becomes

$$\operatorname{div} Tu = 2H \equiv const. \tag{6}$$

in Ω, while (5) remains unchanged.

In (6), the choice of H is not arbitrary. If we integrate (6) over Ω, we find by the divergence theorem

$$2H|\Omega| = \oint_{\Sigma} \nu \cdot Tu\, ds = |\Sigma| \cos\gamma \tag{7}$$

by (5), yielding a value independent of the volume of fluid, so long as Ω is covered.

Let us now choose a subdomain $\Omega^* \subset\subset \Omega$, bounded in part by $\Gamma \subset \Omega$ and by a subarc $\Sigma^* \subset \Sigma$. The divergence theorem now yields

$$2H|\Omega^*| = |\Sigma^*|\cos\gamma + \int_{\Gamma} \nu \cdot Tu\, ds\,. \tag{8}$$

The essential observation that underlies everything that follows is that $|\nu \cdot Tu| < 1$ *for any function u.* Replacing $\nu \cdot Tu$ in (8) by its lower bound -1, we find that the functional

$$\Phi[\Omega^*;\gamma] \equiv |\Gamma| - |\Sigma^*|\cos\gamma + 2H|\Omega^*| > 0 \tag{9}$$

for any non-null choice of $\Omega^* \subset\subset \Omega$.

Consider the particular situation in which Ω contains a corner of opening angle 2α. Data cannot be prescribed at the vertex, but we can exclude that point by a small arc Λ, which is then allowed to shrink to the vertex. Choosing Γ as in Figure 3, we obtain from (9) after a limiting procedure

$$H\, l^2 \sin\alpha\cos\alpha + l\sin\alpha > l\cos\gamma\ .$$

It can be shown that *this holds regardless of behavior at the corner, provided only that* (5) *holds on the* (*open*) *sides in* Σ ; *no growth hypotheses need be introduced.* Letting $l \to 0$, we find $\sin\alpha \geq \cos\gamma$. We have proved:

<u>**Theorem**</u> **1**: *If* Ω *has a corner point of opening angle* 2α, *and if* $\alpha < \left|\frac{\pi}{2} - \gamma\right|$, *then there is no solution to the problem* (4), (5) *in* Ω .

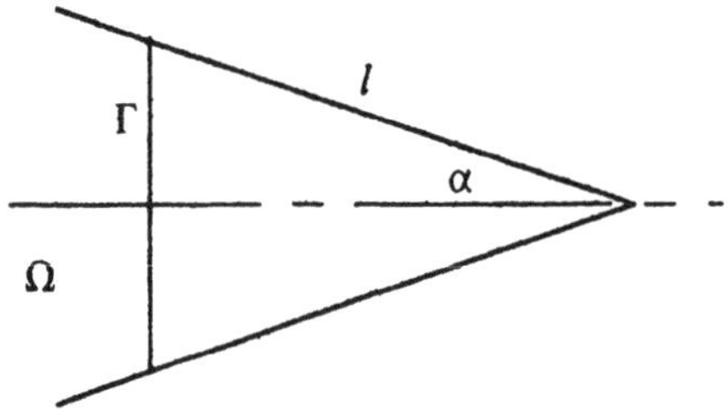

Figure 3: Corner configuration

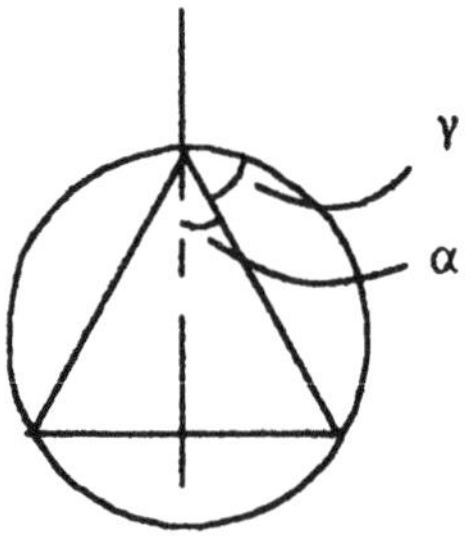

Figure 4: Regular polygon; circumscribed equatorial circle

To put the theorem into context, consider the case in which Ω is a regular polygon Π (for example, the equilateral triangle of Figure 4). A lower hemisphere whose equatorial circle circumscribes Π provides an explicit solution of (4), (5), for which $\gamma = \frac{\pi}{2} - \alpha$. Any γ in the range $\left|\frac{\pi}{2} - \gamma\right| \leq \alpha$ can be obtained by increasing the radius of the hemisphere and/or replacing the lower hemisphere by an upper one.

In fact, for the indicated geometry, the solution u is uniquely determined by the prescribed volume, and is always a spherical cap when it exists. *Thus, the result shows a discontinuous dependence of the solutions on the data* γ. As γ is decreased (or increased) from $\pi/2$, the solution continues to be determined as a spherical cap up to and including a critical value γ_{cr}. But with any change beyond γ_{cr} the solution ceases to exist.

Experiments conducted by W. Masica have shown that the theorem reflects physical reality. He found the predicted spherical cap when $\left|\frac{\pi}{2} - \gamma\right| \leq \alpha$, while for $\left|\frac{\pi}{2} - \gamma\right| > \alpha$ the fluid climbed up the walls at the corners and partly wet the top of the cylinder (or the reverse). This explains the seeming paradox of the theorem; in fact, a surface S does exist as solution of (2), (3) but it cannot be expressed as solution of (4), for the reason that S bends over itself at the corners (Figure 5) and does not cover all of Ω.

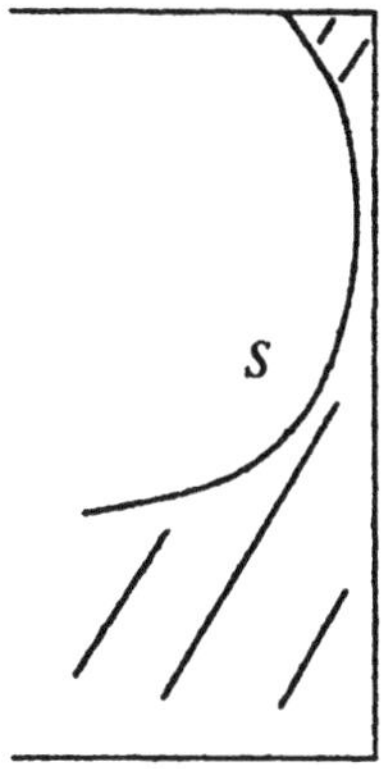

Figure 5: Fluid section at corner

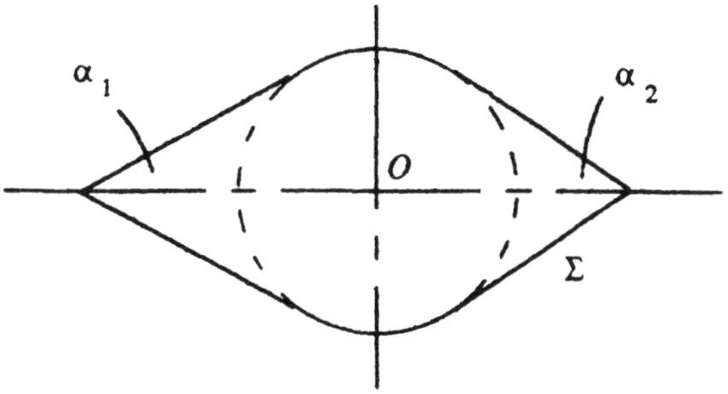

Figure 6: Two corner domain; spherical caps, with concentric circles as equator, meet cylinder over Σ in constant angle

In any event, there is a discontinuity in behavior as the critical contact angle γ_{cr} is crossed, with S changing its character dramatically at the critical value. Clearly the surface forces are the overwhelmingly dominant ones in any such configuration, and it is proposed to use the phenomenon in a space experiment to measure contact angle. Consider the case $0 < \gamma < \frac{\pi}{2}$. An initial guess $\hat{\gamma}$ for γ_{cr} is made, and two angles γ_1, γ_2 are chosen so that $\frac{\pi}{2} > \gamma_1 > \gamma > \gamma_2 > 0$. A cylinder is to be constructed with section Ω as in Figure 6, with $\alpha_1 = \frac{\pi}{2} - \gamma_1$, $\alpha_2 = \frac{\pi}{2} - \gamma_2$. Then lower spherical caps centered at O provide explicit solutions whenever $\frac{\pi}{2} > \gamma \geq \gamma_1$, but no solution exists if $\gamma < \gamma_1$.

If $\gamma_{cr} \geq \gamma_1$, the spherical cap will be observed; if $\gamma_1 > \gamma_{cr} \geq \gamma_2$ the fluid will climb up the smaller corner, and if $\gamma_{cr} < \gamma_2$ the fluid will climb up both walls.

This procedure shows excellent promise for values of γ near $\pi/2$, but can become subject to experimental error for values γ close to zero or π. For such cases, we seek other geometries for which discontinuous (or nearly discontinuous) behavior must occur; since explicit solutions (such as spheres) are available only in limited configurations, we have recourse to general existence (and nonexistence) theorems. We can indicate here only a general outline of the results. For simplicity, we restrict attention to acute angles γ_{cr}.

Theorem 2: *To every Ω, there corresponds a value γ_{cr}, $0 \leq \gamma_{cr} \leq \pi/2$, such that there exists a solution in Ω if $\gamma_{cr} < \gamma < \pi/2$, and there exists no solution when $0 < \gamma < \gamma_{cr}$.*

Any value γ_{cr} in the indicated range can be achieved by particular Ω.

Theorem 3: *If $0 < \gamma_{cr} < \pi/2$ then there exists $\Omega^* \subset \Omega$, $\Omega^* \neq \emptyset, \Omega$, such that $\Phi[\Omega^*, \gamma_{cr}] = 0$. The set Ω^* is bounded in Ω by a finite number of disjoint subarcs Γ of semicircles, each of which meets Σ either in the angle γ_{cr} or at a reentrant corner ($\alpha > \pi/2$), and all of which have radius $R = |\Omega|/(|\Sigma|\cos\gamma_{cr})$.*

Physically, if $\gamma > \gamma_{cr}$ then all solutions of fixed volume remain bounded in Ω, but when γ decreases to γ_{cr} the surfaces become infinitely high throughout Ω^*, and tend to empty the complementary set. In the limit $\gamma = \gamma_{cr}$ we obtain (from a formal mathematical point of view) a *generalized solution in the sense of Miranda* [11], with $u \equiv \infty$ in Ω^*. The arcs Γ are the extremals of a *subsidiary variational problem*, see [12].

For experimental purposes, it is desirable to have a configuration for which Ω^* is large, so as to facilitate observation; in this respect we are motivated by an observation made in [13], for the particular case $\gamma_{cr} = 0$. An explicit example is given there of a domain containing a "strip" of arbitrary length, completely filled by (semicircular) extremals Γ in parallel translation, and for which $\Phi = 0$ for each Γ.

This line of thought is pursued in [14]. With a view toward creating an analogous situation for general γ_{cr}, we seek curves C which make prescribed constant angle γ with a family of circular arcs, of given radius R and in parallel translation. The problem leads to an ordinary differential equation of first order, which can be solved explicitly in the form

$$x + c = \sqrt{R^2 - y^2} + R \sin\gamma \ln \frac{\sqrt{R^2 - y^2}\cos\gamma - y\sin\gamma}{R + y\cos\gamma + \sqrt{R^2 - y^2}\sin\gamma} \tag{10}$$

The integral curves are sketched in Figure 7. We limit attention to those between the horizontal lines (trivial solutions) $y = \pm R\cos\gamma$. Each integral curve is obtained from any given one by rigid horizontal translation. All of them meet each member of the family of translated circles in the given angle γ, and are asymptotic to the two lines as $x \to -\infty$.

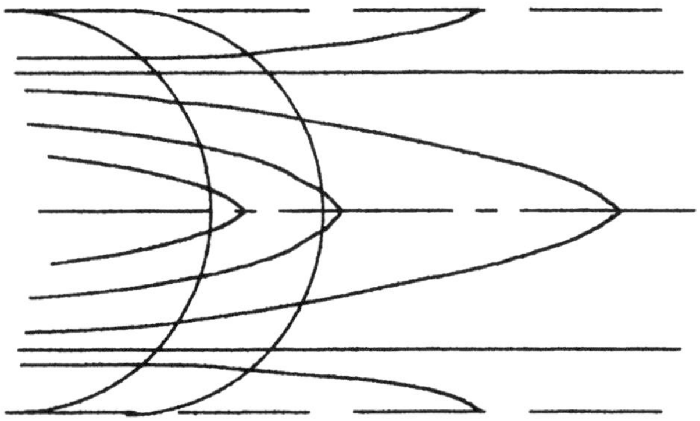

Figure 7: Canonical proboscides; each proboscis meets all translates of the circular arcs in the same angle γ

We pick any particular integral curve C and extend it back as far as desired. We then seek to complete the domain thus created by a "bubble" (Figure 8) of radius ρ chosen so that one will have $R = |\Omega|/(|\Sigma|\cos\gamma_{cr})$. This has been shown empirically to be possible, and in fact in such a way that $R\cos\gamma < \rho < 2R$. With this choice of ρ, we obtain a domain with "proboscis" as large as desired, which is swept out by the extremals Γ of the subsidiary problem as above. From this property we conclude immediately from the variational condition that Φ is independent of the particular extremal for which it is evaluated; since Φ vanishes at the right hand vertex, there must hold $\Phi \equiv 0$ for every extremal.

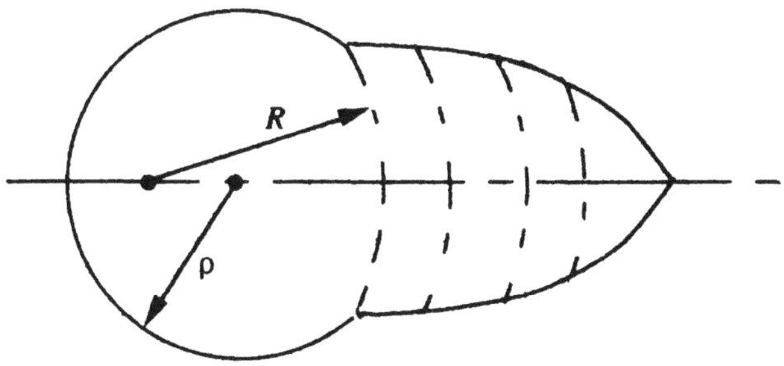

Figure 8: Proboscis domain with bubble, showing extremals Γ

The behavior of solutions as $\gamma \downarrow \gamma_{cr}$ is for this geometry no longer discontinuous as in the case of a wedge; for the proboscis the solution rises unboundedly as γ_{cr} is approached. Numerical studies [15] show however that for a proboscis of reasonable size, the rise is not significant until γ is very close to γ_{cr} , after which it becomes very abrupt. This is the case even for γ_{cr} close to zero, and for a proboscis large enough to require significant fluid displacements as γ_{cr} is approached. *In this way a domain is constructed corresponding to any prescribed* γ_{cr}, *so that a nearly discontinuous behavior will be exhibited at* $\gamma = \gamma_{cr}$ *over a subregion of size large enough to permit clear observation, with little danger of experimental error.*

It must be pointed out that not all details of the formal mathematical analysis are yet complete; however the steps that remain outstanding do not appear to present major obstacles. Numerical investigation of the predicted behavior has already been completed for cases of particular interest [15]. Subject to verification of the last details of the formal analysis, we are prepared to offer the "canonical proboscis" as a workable geometry for precise measurement of contact angles, both large and small, and as a vehicle for general experimental studies relating to the intrinsic significance of the contact angle as a constant of the particular materials.

The work described above was supported in part by grants from the National Science Foundation, and in part by a grant from the National Aeronautics and Space Administration.

References

1. T. Young: *An essay on the cohesion of fluids.* Phil. Trans. Roy. Soc. London, 95 (1805) 55-87

2. P.S. Laplace: *Traité de mécanique céleste*; suppléments au Livre X (1806) in Œuvres Complete Vol. 4, Gauthier-Villars, Paris. See also the annotated English translation by Bowditch (1839), reprinted by Chelsea, New York, 1966

3. C.F. Gauss: *Principia Generalia Theoriae Figurae Fluidorum.* Comment. Soc. Regiae Scient. Gottingensis, Rec. 7 (1830). Reprinted as *Grundlagen einer Theorie der Gestalt von Flüssigkeiten im Zustand des Gleichgewichtes,* in Ostwald's Klassiker der exakten Wissenschaften, vol. 135, W. Engelmann, Leipzig, 1903.

4. R. Finn: "*Equilibrium Capillary Surfaces*". Grundlagen Mathem. Wissen., Springer-Verlag, New York, 1986.

5. T. Young: Reprint (with additional material) of "*An essay on the cohesion of fluids*". In "Miscellaneous Works" (ed. G. Peacock), Vol. 1, John Murray, London (1855) pp. 418-453

6. F. Bashforth and J. G. Adams: *An Attempt to Test the Theories of Capillary Action by Comparing the Theoretical and Measured Forms of Drops of Fluid.* Cambridge Univ. Press, 1883

7. E. B. Dussan V: *On the spreading of liquids on solid surfaces; static and dynamic contact lines.* Ann. Rev. Fluid Mech., 11 (1979) 371-400.

8. J. B. Keller and G. J. Merchant: *Contact angles.* Physics of Fluids, A 4 (1992) 477-485.

9. R. Finn and M. Shinbrot: *The capillary contact angle I: the horizontal plane and stick-slip motion.* J. Math. Anal. Appl. 123 (1987) 1-17.

10. R. Finn and M. Shinbrot: *The capillary contact angle II: the inclined plane.* Mathem. methods in Appl. Sci. 10 (1988) 165-196.

11. M. Miranda: *Superfici minime illimitate.* Ann. Scuola Norm. Sup. Pisa, 4 (1977) 313-322

12. R. Finn: *A subsidiary variational problem and existence criteria for capillary surfaces.* J. Reine Angew. Math. 353 (1984) 196-214.

13. P. Concus and R. Finn: *Continuous and discontinuous disappearance of capillary surfaces.* In "*Variational Methods for Free Surface Interfaces*", Springer-Verlag, New York, 1987.

14. B. S. Fischer and R. Finn: *Existence theorems and measurement of the capillary contact angle.* Zeit. Anal. Anw., to appear.

15. P. Concus, R. Finn and F. Zabihi: *On canonical cylinder sections for accurate determination of contact angle in microgravity.* Lawrence Berkeley Lab Report LBL-32279, (1992).

Third International Colloquium on Differential Equations pp. 63-76 (1993)
D. Bainov and V. Covachev (Eds)

Kernel Asymptotics of Exotic Second-Order Operators

Stephen A. Fulling

Mathematics Department, Texas A&M University, College Station, Texas, 77843-3368 U.S.A.

ABSTRACT

The Navier–Lamé operator of classical elasticity, $\mu\Delta\mathbf{v} + (\lambda+\mu)\nabla(\nabla\cdot\mathbf{v})$, and the operator $a^2 d\delta + b^2\delta d$ on differential forms are the simplest examples of differential operators whose second-order terms involve a coupling among the components of a vector-valued function. I show how the heat-kernel expansion of such operators can be computed by the intrinsic symbolic calculus of pseudodifferential operators with computer assistance.

Physics, old and new. Let Ω be a region in $\mathbf{R}^3$, and $\mathbf{v}(t,\mathbf{x})\colon \mathbf{R}\times\Omega \to \mathbf{R}^3$ a vector field. Elastic waves in a solid are described by the *Navier equations*,

$$\begin{aligned}\frac{\partial^2\mathbf{v}}{\partial t^2} &= \mu\Delta\mathbf{v} + (\lambda+\mu)\nabla(\nabla\cdot\mathbf{v}) \\ &= (\lambda+2\mu)\nabla(\nabla\cdot\mathbf{v}) - \mu\nabla\times(\nabla\times\mathbf{v}),\end{aligned} \tag{1}$$

where the *Lamé constants*, λ and μ, characterize the material [7, 25]. This equation is distinguished from the other second-order PDEs of mathematical physics by the fact that its second-order terms are not merely the Laplacian, or even the Laplace–Beltrami operator of a nontrivial metric. Since both the independent and the dependent variable reside in $\mathbf{R}^3$, it is possible for them to be "tangled together" in the algebraic structure of the second-order derivative term. The result is an "exotic" differential operator.

Early in the twentieth century, Hermann Weyl investigated the asymptotic spectral behavior of this elastic operator [32–33, 2]. He found that for suitable boundary conditions, all the normal modes could be classified as either *longitudinal* waves, which can be constructed as the gradients of scalar functions ($\mathbf{v} = \nabla\phi$), or *transverse* waves, which satisfy the same divergence condition as electric fields ($\nabla\cdot\mathbf{v} = 0$). Thus the problem decomposes into two previously solved problems, the scalar and the electromagnetic one.

This observation is a special case of the *Hodge decomposition* in differential geometry. Vector fields on a Riemannian manifold are naturally identified with 1-forms. Let d be the operation of exterior differentiation, mapping k-forms into $(k+1)$-forms, and δ be the negative of its adjoint, mapping in the opposite direction. Then the Navier–Lamé operator (1) is of the form $B_1\delta d + B_2 d\delta$ for certain constants B_j, with $k = 1$. Longitudinal and transverse modes belong to the null-spaces of d and δ, respectively.

In recent years exotic operators of a similar sort have been encountered by physicists constructing quantum theories of gravity [3, 4, 30]. They are associated with so-called "ghost" degrees of freedom in theories where the gravitational field obeys fourth-order equations of motion. For example, Barth and Christensen [3] define an operator F by

$$Fv_\alpha = \Delta v_\alpha + \nabla^\beta(\nabla_\alpha v_\beta) - 2\eta\nabla_\alpha\nabla^\beta v_\beta\,, \tag{2}$$

where Δ is the (Bochner) Laplace–Beltrami operator and ∇ the covariant derivative relative to a semi-Riemannian metric on a four-dimensional manifold; η is a parameter formed out of the coupling constants of the gravitational theory. Commutation of derivatives in the middle term shows F to be a curved-space generalization of the Navier–Lamé operator (1), plus a zeroth-order term built out of the curvature tensor of the manifold. Operators of this type, acting on k-forms, have been called *generalized Ahlfors Laplacians* [6]. Some applications of them within differential geometry are cited in [21].

Exotic second-order terms interfere with the standard methods of calculating asymptotic expansions of Green functions, effective Lagrangians, and so on. These are central tools of the quantum gravity theorists (and of mathematicians studying index theorems or inverse problems, and of physicists studying atomic nuclei or the thermodynamics of small grains of material). The general setting (restricted to the second-order case) is an operator

$$H = A^{\mu\nu}(x)\nabla_\mu\nabla_\nu + B^\mu(x)\nabla_\mu + C(x) \tag{3}$$

acting on sections of a vector bundle over a [semi-]Riemannian manifold M with metric tensor $g_{\mu\nu}$. (Here $x \in M$ is the independent variable, and the summation convention over repeated tensor indices is in force.) Locally each coefficient $A^{\mu\nu}$, B^μ, or C can be represented as a matrix (with respect to a local basis for the fiber space at x of the bundle). In nonexotic operators (which have come to be called "minimal" operators in the physics literature) $A^{\mu\nu}$ is proportional to $g^{\mu\nu}$ times the identity matrix, or $[A^{\mu\nu}]^a{}_b = \text{const.}\, g^{\mu\nu}\delta^a_b$. For the simplest exotic operators, such as (1) and (2), the bundle is the tangent (or cotangent) bundle of M, and

$$[A^{\mu\nu}]^\beta{}_\alpha = C_1\, g^{\mu\nu}\delta^\beta_\alpha + C_2\,\tfrac{1}{2}(g^{\mu\beta}\delta^\nu_\alpha + g^{\nu\beta}\delta^\mu_\alpha); \tag{4}$$

the last term "tangles" directions in the bundle with directions in the manifold, producing the distinctive "grad div" structure.

Three calculational methods. Let $K(t,x,y)$ be the integral kernel of the operator e^{-tH} — equivalently, the Green function solving the initial value problem for the "heat equation" $\partial\mathbf{v}/\partial t = -H\mathbf{v}$. (For greater precision, suppose that the differential expression H described earlier is formally self-adjoint, and understand H in the present discussion to be any self-adjoint realization of it.) It is well known that as $t \downarrow 0$ the diagonal value of the heat kernel has an asymptotic expansion of the form

$$K(t,x,x) \sim (4\pi t)^{-m/2}\sum_{n=0}^{\infty} t^n a_n(x). \tag{5}$$

This is true for exotic operators as well as minimal ones [22, 31]. (Here m is the dimension of the base manifold M. Note that K and a_n are matrix-valued (sections of the bundle of endomorphisms of the fiber bundle).) From (5) many other expansions of interest can be derived. (This "local" expansion merely scratches the surface of the geometrical information contained in the asymptotics of Green functions and spectra. Much more delicate analysis is needed to extract the effects of boundaries, closed geodesics, etc.; those matters are beyond the scope of this presentation.)

For nonexotic second-order operators, (5) has an off-diagonal generalization,

$$K(t,x,y) \sim e^{-d(x,y)^2/4t}(4\pi t)^{-m/2}\sum_{n=0}^{\infty} t^n a_n(x,y), \tag{6}$$

where d is the semi-Riemannian geodesic distance function. Substitution into the differential equation defining K then yields recursion relations that can be solved (laboriously) for the a_n [e.g., 28, 9]. (This process is known in the physics literature as the *Schwinger–DeWitt expansion.* In mathematics the series is associated with Hadamard and Minakshisundaram.) However, for fourth-order operators [26, 8] and exotic second-order operators such a simple factorization does not exist, and this classical method does not provide a recursive algorithm.

A second approach to finding the $a_n(x)$, or at least the integrals of their traces over M, is some variant of this: At each order n, a_n must be a linear combination of a finite list of allowable invariant objects built from the coefficient tensors in H, the curvature tensor, etc.; the problem is to determine its numerical coefficients. Certain relations among the coefficients can be deduced from general principles, such as how the heat kernel of an operator on a product manifold is related to the heat kernels of the factors. Other relations can be found by looking at special cases for which the answer can be calculated easily. With luck one can compile enough information to fix the coefficients uniquely. This strategy was employed in the famous paper of McKean and Singer [27] and has been extensively developed by Peter Gilkey [e.g., 19]. Its great advantage is that, when it works, it provides results very efficiently, while providing some interesting geometrical insights. Its disadvantage is that it is not an algorithm; to push the method forward to a higher order or a more general class of operator, additional mathematical creativity is always required.

Gilkey and Branson [21] in this way found some important information about the integrated trace of the heat kernel of the natural exotic operators on differential forms on a compact Riemannian manifold without boundary. (On such a manifold (5) is uniformly valid and can be integrated, giving

$$\begin{aligned}\int_M \operatorname{tr} H(t,x,x)\,dx &\sim (4\pi t)^{-m/2}\sum_{n=0}^{\infty} t^n a_n(H)\\ &= \sum_{\nu}^{\infty} e^{-t\lambda_\nu} \equiv \operatorname{Tr} e^{-tH},\end{aligned} \tag{7}$$

$$a_n(H) \equiv \int_M \operatorname{tr} a_n(x)\,dx, \tag{8}$$

where the λ_ν are the eigenvalues of H.) In the role of H consider

$$D = a^2 d\delta + b^2 \delta d \tag{9}$$

operating on k-forms ($C^\infty(\Lambda^k(M))$). If $a = b$, D is proportional to the *DeRham Laplacian* on forms,

$$\Delta_k = d\delta + \delta d \qquad \text{on } C^\infty(\Lambda^k), \tag{10}$$

which is a nonexotic operator. The $a_n(\Delta_j)$ for $0 \le j \le m$ may thus be regarded as "known". (In fact, $-\Delta_j$ differs only by a zeroth-order term — the *Weitzenböck endomorphism* — from the *Bochner Laplacian*, $\Delta \equiv g^{\mu\nu}\nabla_\mu\nabla_\nu$, and $a_n(x)$ for such an operator has been determined up through $n = 4$ [1]. Formulas for $a_0(\Delta_j)$ and $a_1(\Delta_j)$ are provided in Theorem 1.1 of [21].) As previously remarked, the Hodge decomposition cuts k-forms into "longitudinal" and "transverse" parts; D effectively acts on these two spaces separately, and hence a "telescoping" calculation [6, 21] expresses the trace of e^{-tD} in terms of the traces of $e^{-t\Delta_j}$ for $j \le k$, in a reminiscence of Weyl's classic treatment of the elasticity problem.

THEOREM 1. *With the definitions (7) [or (8)], (9), and (10), one has*

$$a_n(D) = b^{2n-m} a_n(\Delta_k) + \left(b^{2n-m} - a^{2n-m}\right) \sum_{j<k} (-1)^{k-j} a_n(\Delta_j). \tag{11}$$

An interesting broader class of exotic operators (including (2), for instance) comprises those of the form

$$H = a^2 d\delta + b^2 \delta d - E(x), \tag{12}$$

where E is a zeroth-order operator (a matrix-valued function). In [21], $a_0(H)$ and $a_1(H)$ were determined for such operators. (See [6] for an extension to manifolds with boundary.)

THEOREM 2. *For the operator (12) on k-forms, one has (in terms of the quantities in Theorem 1)*

$$a_0(H) = a_0(D), \tag{13}$$

$$a_1(H) = a_1(D) + \frac{a_0(D)}{\binom{m}{k} \int_M 1\, dx} \int_M \operatorname{tr} E(x)\, dx. \tag{14}$$

(The integral in the denominator in (14) is simply the volume of M, which appears as a factor in $a_0(D)$.)

PROOF: The only invariants eligible to appear in a_1 are the curvature scalar R (already contained in $a_1(D)$) and $\operatorname{tr} E$. Therefore, it suffices to determine the coefficient of the trace term by considering the case $E =$ identity operator. For that case we have in (7)

$$\operatorname{Tr} e^{-tH} = e^t \operatorname{Tr} e^{-tD}$$

and hence

$$a_0(H) + t a_1(H) + \cdots = a_0(D) + t[a_1(D) + a_0(D)] + \cdots. \tag{15}$$

For the identity we also have

$$\operatorname{tr} E = \dim \Lambda^k(M) = \binom{m}{k},$$

so the $O(t)$ part of (15) can be written in the generic form (14).

This argument does not generalize in any obvious way to give $a_n(H)$ for $n \geq 2$. (For example, $a_2(H)$ contains both $\operatorname{tr}(E^2)$ and $(\operatorname{tr} E)^2$ (not to mention terms bilinear in E and in curvature), and these are not separated by the identity operator.) Moreover, formulas for the global quantities $a_n(H)$ do not give complete information about the local quantities $a_n(x)$. For one thing, any exact divergence, such as $\Delta_0 R$, integrates to zero. For another, $[a_1(x)]^\alpha{}_\beta$ may contain both terms proportional to $[E(x)]^\alpha{}_\beta$ and terms proportional to $\operatorname{tr} E(x)\, \delta^\alpha_\beta$, and these are no longer independent after the trace is taken in (8) and (14). (Traces of the coefficient tensors of H do not appear in the untraced $a_n(x)$ for ordinary operators H, but if H is exotic, they do.)

A third way to calculate $a_n(x)$ is based on the calculus of pseudodifferential operators [31, 22, 18, 34, 30, 16, 23, 24]. It has broader validity than the two traditional methods; it offers in principle a complete solution of the problem for operators of the class (12). (For more general exotic (3) or higher-order operators, integrals may be encountered that cannot be evaluated in closed form.) Unfortunately, it is much less efficient than the other methods. An attack upon its computational complexity is the main subject of the present work.

The symbol. In the *intrinsic pseudodifferential calculus* [5, 34, 11, 16] each covariant derivative, ∇, in an operator such as (3) is represented by a Fourier (cotangent bundle) variable, $i\xi$. Thus the operator on $C^\infty(\Lambda^1)$

$$H_0 \equiv -\Delta - \frac{c}{2}\left(\nabla_\beta \nabla^\alpha + \nabla^\alpha \nabla_\beta\right) \qquad (\Delta \equiv g^{\mu\nu}\nabla_\mu\nabla_\nu) \tag{16}$$

has the intrinsic symbol

$$\operatorname{Sy}(H_0) = |\xi|^2 + c\,\xi \otimes \xi. \tag{17}$$

By definition, an *exotic* operator is one with *nonscalar principal symbol*; that is, the terms in the symbol of highest degree in ξ are not just a multiple of the identity matrix. We may think of H_0 as operating either on vector fields or on one-forms; to match up with physics literature, I shall consider vector fields. Then the coefficient tensor in (16) and (17) is

$$[A^{\mu\nu}]^\beta{}_\alpha = -g^{\mu\nu}\delta^\beta_\alpha - \frac{c}{2}\left(g^{\mu\beta}\delta^\nu_\alpha + g^{\nu\beta}\delta^\mu_\alpha\right). \tag{18}$$

Here all the indices are Greek, since they refer to the same vector bundle (or its dual); their place of origin in the symbol is momentarily preserved by the distinction between the beginning and middle of the alphabet.

We consider the class of operators

$$H = b^2 H_0 + V(x). \tag{19}$$

(Here b^2 and c are constants, obeying sign constraints to be discussed presently, and V is a C^∞ endomorphism-valued function, called the *potential* in analogy with quantum mechanics.) Equivalently, H is of the form (12) with

$$-E^\beta{}_\alpha = V^\beta{}_\alpha - b^2\left(1+\frac{c}{2}\right)R^\beta{}_\alpha \tag{20}$$

and

$$a^2 = b^2(c+1), \qquad c = \frac{a^2-b^2}{b^2}\,. \tag{21}$$

The operator can be parametrized by a and b, to exploit the Hodge decomposition, or by b and c, so that c is the magnitude of the exotic term and b is merely a scale factor. (In [24], V is called X and c is called $-a$. The terms in (20) proportional to the Ricci tensor $R^\beta{}_\alpha$ are the Weitzenböck operator for $k=1$ and a similar contribution from the desymmetrization of the exotic term in (16).)

From the second-order derivative terms one reads off the principal symbol of H as

$$\begin{aligned} +b^2(|\xi|^2 + c\,\xi\otimes\xi) &= b^2|\xi|^2 + (a^2-b^2)\,\xi\otimes\xi \\ &= a^2\mathrm{ext}_\xi\mathrm{int}_\xi + b^2\mathrm{int}_\xi\mathrm{ext}_\xi\,, \end{aligned} \tag{22}$$

where int and ext are the operators of interior and exterior multiplication on forms. (Here $\xi\otimes\xi$ has the matrix

$$\xi^\beta\xi_\alpha$$

in a conventional local basis.) Under the assumption that the metric is positive definite and $b^2>0$, the condition that both eigenvalues be nonnegative, so that the heat kernel exists, is

$$c \ge -1, \quad \text{or} \quad a^2 \ge 0. \tag{23}$$

By convention, $b>0$ and $a\ge 0$.

In the calculation a major role is played by the eigenvalues and eigenprojections of the matrix (22). They are

$$\lambda_1 = a^2|\xi|^2 = b^2(c+1)|\xi|^2, \qquad P_1 = \frac{\xi\otimes\xi}{|\xi|^2} \quad (\text{along } \xi), \tag{24a}$$

$$\lambda_2 = b^2|\xi|^2, \qquad P_2 = \frac{|\xi|^2 - \xi\otimes\xi}{|\xi|^2} \quad (\perp \text{ to } \xi). \tag{24b}$$

In the notation of [16] we have

$$b_0(\lambda) \equiv [\mathrm{Sy}(b^2H_0)-\lambda]^{-1} = -\sum_{j=1}^{2}\frac{P_j}{\lambda-\lambda_j}\,. \tag{25}$$

The first step in calculating the heat kernel of H is to find the symbol of the resolvent operator, $(H-\lambda)^{-1}$, following [34] and [16]. For background reading on the resolvent symbol and the heat kernel, I highly recommend the expositions of Gilkey [17, 18, 20]. (They, however, deal with the *conventional* pseudodifferential calculus, where ξ is a literal Fourier variable corresponding to conventional partial differentiation. The basic formulas of that approach are simpler than those of the intrinsic formalism, but extra work is needed at the end to express the results in coordinate-independent geometrical form.) The series of conference reports [14–15] provides a brief introduction to and summary of [16].

THEOREM 3. *The intrinsic symbol of the resolvent parametrix of an elliptic linear differential operator has an asymptotic expansion*

$$Sy[(H-\lambda)^{-1}] \sim \sum_{s\geq 0} b_s(x,\xi,\lambda), \tag{26}$$

where b_0, as in (25), is the local resolvent of the principal symbol, and the higher b_s are given by an explicit formula [15, 16] involving very complicated index contractions and summations over multiindices. The formula for b_2 contains 40 terms for a generic operator, but only 5 in the present case (19):

$$\begin{aligned} b_2 = &- b_0 V b_0 \\ &- 4 b_0 A^{\alpha\mu} b_0 A^{\beta\nu} (\nabla_\mu \nabla_\nu I) b_0\, \xi_\alpha \xi_\beta \\ &- 2 b_0 A^{\alpha\mu} b_0 A^{\nu\rho} (\nabla_\mu \nabla_\nu \nabla_\rho \Phi^\beta) b_0\, \xi_\alpha \xi_\beta \\ &- 2 b_0 A^{\mu\nu} b_0 A^{\alpha\rho} (\nabla_\mu \nabla_\nu \nabla_\rho \Phi^\beta) b_0\, \xi_\alpha \xi_\beta \\ &- 8 b_0 A^{\alpha\mu} b_0 A^{\beta\nu} b_0 A^{\gamma\rho} (\nabla_{(\mu} \nabla_{\nu)} \nabla_\rho \Phi^\delta) b_0\, \xi_\alpha \xi_\beta \xi_\gamma \xi_\delta \,. \end{aligned} \tag{27}$$

[Here a matrix multiplication is implied, and hence the fiber-bundle indices, including the last two indices in (18), are suppressed. I is the parallel transport operator in the tangent bundle, called τ^E in [16]. See (A.11–13) of [24]; their W will be 0 for us. $\Phi(x,y)$ is the tangent vector to the geodesic joining x and y (see [16], Remark 2.5). The important thing is that in the present case all the covariant derivatives of I and Φ, evaluated at $y = x$ as tacitly implied in (27), can be recursively calculated in terms of the Riemann tensor [9, 10, 12].] *More generally, a term in b_s has the form*

$$(\text{coefficient})\, b_0 T_U b_0 T_{U-1} \cdots b_0 T_1 b_0\, \xi^{\otimes(2U-s)}, \tag{28}$$

where the T_u are tensors built out of A, V, Φ, and I and their covariant derivatives, with indices absorbing the ξ factors by contraction. (Each T is linear in (A,V).) U ranges from $U_0 \equiv -[-s/2] = \lceil s/2 \rceil$ to $2s$.

For our special operator, the covariant derivatives of A vanish, and $1 \leq U \leq 4$ for $s = 2$. To obtain the diagonal value of the heat kernel we follow Widom [34], pp. 59–61.

THEOREM 4. *In an orthonormal frame at x we have*

$$K(t,x,x) \sim \sum_{s=0}^{\infty} K_s(x), \tag{29}$$

$$K_s(x) = (2\pi)^{-m} \int_{\mathbf{R}^m} d^m\xi \left(\frac{-1}{2\pi i}\right) \int_\Gamma d\lambda\, e^{-t\lambda} b_s(x,\xi,\lambda), \tag{30}$$

where Γ *surrounds* λ_1 *and* λ_2 *in the positive sense.* K_s *will be 0 for* s *odd.* [In the contrary case, K_{2n} equals $(4\pi t)^{-m/2}\, t^n a_n$ in our earlier notation (5).] *Thus the contribution of the term (28) is*

$$t^{(s-m)/2}(\text{coefficient})(-1)^U(2\pi)^{-m}\int d^m\eta \sum_{i_0=1}^{2}\cdots\sum_{i_U=1}^{2} P_{i_U}T_U\cdots T_1P_{i_0} \otimes \eta^{\otimes(2U-s)}\frac{1}{2\pi i}\int_\Gamma d\mu\, e^{-\mu}(\mu-\mu_1)^{-M_1}(\mu-\mu_2)^{-M_2}, \tag{31}$$

where

$$M_j \equiv \textit{cardinality of } \{i_\iota : i_\iota = j\} \qquad (M_1 + M_2 = U + 1), \tag{32}$$

$$\mu_1 \equiv a^2|\eta|^2, \qquad \mu_2 = b^2|\eta|^2. \tag{33}$$

To get to (31) one makes the substitutions

$$\eta = t^{1/2}\xi, \qquad \mu = t\lambda$$

and renames $t\Gamma$ as Γ. Whenever T_u equals A, another sum from 1 to 2 can be introduced into (31), corresponding to the two terms in (18); furthermore, P_2 splits into two terms (24b), so the i sums are effectively over three values. This implies that each term in b_{2n} gives rise to a rather large number of terms in a_n — namely, $3^{U+1}2^{U-v}$, where v is the number of occurrences of the potential V in the term (28). Thus from (27) for b_2 we will get (in a_1) 9 terms involving V and $108 + 108 + 108 + 648 = 972$ other terms, which are all linear in the Ricci tensor. The number of terms in a_2 is over a million! These terms are highly redundant, in the senses that many of them are manifestly proportional, quite a few turn out to be zero, and they are all linear combinations of a small number of linearly independent objects. For example, a_1 must simplify to a sum of 5 terms, proportional to V, its transpose, its trace (times the identity), the Ricci tensor, and its trace (the curvature scalar R).

In what follows I shall demonstrate that all the integrations in (31) can be performed in closed form, so that in principle $a_n(x)$ has been expressed in terms of elementary functions (and local, polynomial functionals of the coefficient tensors). Furthermore, I shall show that modern computer technology makes practical the actual calculation of a_1 and probably a_2 by this method.

Integrals. In evaluating formula (31), one encounters three kinds of integrals:

1. *Cauchy integrals* over the spectral parameter, μ.
2. *Gaussian integrals* over the radial coordinate in Fourier space, $|\eta|$.
3. *Angular integrals* over the unit sphere in Fourier space.

THEOREM 5. *Define*

$$F_{M_1M_2}(\mu_1,\mu_2) \equiv \frac{1}{2\pi i}\int_\Gamma e^{-\lambda}(\lambda-\mu_1)^{-M_1}(\lambda-\mu_2)^{-M_2}\,d\lambda, \tag{34}$$

where

$$\mu_1 = a^2|\eta|^2, \qquad \mu_2 = b^2|\eta|^2, \qquad c \equiv \frac{a^2 - b^2}{b^2} \neq 0.$$

Then

$$\begin{aligned} F_{M_1M_2}(\mu_1,\mu_2) = &\sum_{J=0}^{M_1-1} \frac{(-1)^{M_1-1}(M_2+J-1)!}{(M_2-1)!\,(M_1-J-1)!\,J!}(b^2c)^{-M_2-J}|\eta|^{-2(M_2+J)}e^{-a^2|\eta|^2} \\ &-\sum_{J=0}^{M_2-1} \frac{(-1)^{M_2-M_1-J}(M_1+J-1)!}{(M_1-1)!\,(M_2-J-1)!\,J!}(b^2c)^{-M_1-J}|\eta|^{-2(M_1+J)}e^{-b^2|\eta|^2}, \end{aligned} \tag{35}$$

$$F_{M_10}(\mu_1,\mu_2) = \frac{(-1)^{M_1-1}}{(M_1-1)!}e^{-a^2|\eta|^2}, \qquad F_{0M_2}(\mu_1,\mu_2) = \frac{(-1)^{M_2-1}}{(M_2-1)!}e^{-b^2|\eta|^2}. \tag{36}$$

The proof is a straightforward application of Cauchy's integral formula. Contrary to appearance, the functions in (35) are guaranteed to be nonsingular as $\eta \to 0$.

Henceforth I write $F_{M_1M_2}(a,b)$ instead of $F_{M_1M_2}(\mu_1,\mu_2)$ and r in place of $|\eta|$.

We shall dispose of the angular integrals quickly, since a similar problem has been discussed in depth in the appendix of [8].

THEOREM 6. *Consider an integral of the form*

$$\mathcal{I}_F \equiv \int_{\mathbf{R}^m} d^m\eta\, F(|\eta|)\,\eta^\alpha, \tag{37}$$

where α is a multiindex, corresponding to a string of tensor indices $\mu_1, \mu_2, \dots, \mu_{|\alpha|}$ (unordered, but not necessarily distinct). $\mathcal{I}_F$ equals 0 unless all components of α are even. If $\alpha = 2\beta$, then

$$\mathcal{I}_F = \frac{I_F}{(2|\beta|-1)!!}\,g_{2\beta}, \tag{38}$$

where

$$I_F \equiv \int_{\mathbf{R}^m} d^m\eta\, F(|\eta|)\,{\eta_m}^{2|\beta|}, \tag{39}$$

and

$$g_{2\beta} \equiv g_{\mu_1\mu_2}g_{\mu_3\mu_4}\cdots + g_{\mu_1\mu_3}g_{\mu_2\mu_4}\cdots + \cdots \tag{40}$$

involves a sum over the $(2|\beta|-1)!!$ essentially distinct permutations of the indices. (As a factor in a scalar integrand, this tensor produces a contraction over all possible pairings of the indices in α.) Furthermore,

$$I_F = \pi^{m/2}\,\frac{2^{1-|\beta|}(2|\beta|-1)!!}{\Gamma(\frac{m}{2}+|\beta|)}\int_0^\infty r^{m+2|\beta|-1}F(r)\,dr. \tag{41}$$

(The numerical factors in (41) are the product of $\int_0^\pi \cos^{2|\beta|}\theta \sin^{m-2}\theta\,d\theta$ and the volume of the unit $(m-2)$-sphere.)

We turn to the final step of integration: Theorems 5 and 6 reduce (31) to integrals of the form

$$\int_0^\infty r^{m-1+2U-s}\,F_{M_1M_2}(r)\,dr. \tag{42}$$

All such integrals can be evaluated by the formula

$$\int_0^\infty e^{-qr^2} r^p \, dr = \frac{1}{2} q^{-(p+1)/2} \Gamma\left(\frac{p+1}{2}\right). \tag{43}$$

Gusynin et al. [24] evaluated the Cauchy, and the resulting radial, integrals in terms of hypergeometric functions. In this way they succeeded in calculating a_0 and a_1 by hand. From the foregoing it is clear that all hypergeometric functions appearing in their results can be expressed in elementary functions, perhaps at the cost of less compact formulas.

The leading term. In contrast to the usual case, for an exotic operator even a_0 is nontrivial. Evaluating (31) with (28) collapsed to b_0 , one gets

$$\begin{aligned} a_0(x) &= b^{-m} + \frac{a^{-m} - b^{-m}}{m} \\ &= \frac{1}{b^m m}\left[m - 1 + (c+1)^{-m/2}\right]. \end{aligned} \tag{44}$$

(An identity matrix is implicit here.) When $a = b$ (or $c = 0$) this reduces to the well known "minimal" result, $a_0 = b^{-m}$ (that is, $K_0 = (4\pi b^2 t)^{-m/2}$). When we trace (44) and integrate it over the manifold, we get

$$a_0(H) = \left[m b^{-m} + a^{-m} - b^{-m}\right] \int_M 1 \, dx, \tag{45}$$

in accordance with Theorems 1 and 2.

Computerization. In pursuit of a_1 and a_2 a sequence of computer programs has been written. I have found, here as elsewhere [13], that "homegrown" programs are most effective in generating the large number of terms in asymptotic calculations, but that commercial, general-purpose computer algebra programs [35, 29] are essential for combining like terms and simplifying the output.

1. To implement the result of [16], I wrote a program in C to list the terms in each order, b_s , of the resolvent symbol. This program prints the values of the multiindices characterizing each term (28) and calculates the numerical coefficient. This program is quite general: for example, it could be used to generate the terms in the resolvent of a fourth-order nonexotic operator [23], and it is applicable to the conventional as well as the intrinsic pseudodifferential calculus.

2. That program acts as the basic engine in another program that lists the terms in the heat kernel of H (see (31) and following discussion). The new program, implemented as a subroutine of the other one and constantly revised as the need arises, calculates the new numerical factors accumulated in (31), (38), and (41) and outputs each term in a form suitable as input for *MathTensor*. (It is here that the structure of *tensor contraction*, described verbally in [16] and glossed over in Theorem 3, must be concretely implemented.)

3. A *Mathematica* package was written to evaluate the Cauchy and radial integrals, following (34)–(36) and (42)–(43).

4. An additional *Mathematica* input file is needed to define for *MathTensor* the various tensors appearing in the T_u of (31), to simplify Γ functions, etc. For large $|\beta|$, creating the expression (40) is itself a nontrivial programming task, and digesting it the most time-consuming part of the later computation.

5. *Mathematica/MathTensor* is used interactively to work the various contributions to a_n into a usable final form.

Since b in (19) enters each category of terms only as a scale factor, it is convenient to set $b = 1$ in the calculation; the appropriate power of b is easily restored at the end. The results are most elegantly stated as rational functions in c and

$$\tilde{a} \equiv a/b = \sqrt{c+1}. \tag{46}$$

With this technology, the calculation of the terms in a_1 involving the potential V is fairly simple. We have

$$\begin{aligned} a_1^{(V)}(x)^\beta{}_\alpha &= \frac{\tilde{a}^m[c(2-m)+4] - c(m+2) - 4}{b^m \tilde{a}^m c\, m(m^2-4)} [V_\alpha{}^\beta + V^\mu{}_\mu \delta^\beta_\alpha] \\ &+ \frac{-\tilde{a}^m[c(m^3-2m^2-3m+6)+4(m+1)] + 3c(m+2) + 4(m+1)}{b^m \tilde{a}^m c\, m(m^2-4)} V^\beta{}_\alpha . \end{aligned} \tag{47}$$

(Even if H is formally self-adjoint, forcing V to be Hermitian, V may still be a complex matrix; hence V and its transpose are independent objects.) Taking the trace yields

$$a_1^{(V)}(x)^\mu{}_\mu = -\frac{m - 1 + \tilde{a}^{-m}}{b^m m} V^\mu{}_\mu \tag{48}$$

in agreement with [21] (cf. (14) and (44)).

The curvature terms in a_1 proved to be much more involved, and the computation was barely finished in time for this conference. I find

$$\begin{aligned} a_1^{(R)}(x)^\beta{}_\alpha &= \frac{\begin{array}{c}\tilde{a}^m[3c^2(m-2)+c(m^3-m^2-24)-24] \\ + c^2(m^2+5m+6) + c(m^2+8m+24) + 24\end{array}}{6b^{m-2}\tilde{a}^m c\, m(m^2-4)} R\delta^\beta_\alpha \\ &+ \frac{\begin{array}{c}\tilde{a}^m[3c^2(m^3-2m^2-2m+4)+2c(-5m^2+14m+12)+24m] \\ - 2c^2(m^2+5m+6) - 2c(m^2+14m+12) - 24m\end{array}}{6b^{m-2}\tilde{a}^m c\, m(m^2-4)} R^\beta{}_\alpha . \end{aligned} \tag{49}$$

The trace is

$$a_1^{(R)}(x)^\mu{}_\mu = \frac{\tilde{a}^m[3c(m-1)+m^2-m-6] + c(m+3) + m + 6}{6b^{m-2}\tilde{a}^m m} R. \tag{50}$$

However, the result is much simpler for the operator $D = a^2 d\delta + b^2 \delta d$ defined in (9) (the operator that is simplest and most natural from the point of view of the geometer, as opposed to the Fourier analyst). From (20), (47), and (49) one finds

$$\begin{aligned} a_1^{(D)}(x)^\beta{}_\alpha = {} & \frac{\tilde{a}^m(m-3)+c+1}{6b^{m-2}\tilde{a}^m(m-2)} R\delta^\beta_\alpha \\ & + \frac{\tilde{a}^m(-3m+7)-c-1}{3b^{m-2}\tilde{a}^m(m-2)} R^\beta{}_\alpha . \end{aligned} \tag{51}$$

Its trace is

$$a_1^{(D)}(x)^\mu{}_\mu = \tfrac{1}{6} b^{2-m}[m-7+\tilde{a}^{-m}(c+1)]\, R, \tag{52}$$

which is the prediction of Theorem 1.

MAIN THEOREM. [Formulas (44), (47), (49), and (51) are the main results. The notation is defined in (5), (16), and (19). Results (44) and (47) agree with [24], where a different but related method was employed; comparison of (49) with [24] has not yet been completed.]

After debugging, the *Mathematica* calculation of (49) requires two days of operation of a Sun 3/60 workstation. (Direct calculation of (50) is about 5 times faster. The execution time of the preparatory C program is negligible.) An attack on a_2, therefore, will require either a supercomputer or a more efficient algorithm. Experience gained in this computation indicates several ways in which greater efficiency can be achieved. In particular, the many terms that turn out to be zero tend to take more time (individually!) to compute than do the interesting terms; it is possible to state general principles for discarding some such terms a priori.

Concluding remarks. This work has provided valuable experience in the use of symbolic computation on problems of this nature. Efficiency is strongly dependent on the organization of the calculation, in ways that are not always obvious at the start. The result, and the hoped-for future calculation of a_2, have applications in quantum gravity, and one may hope that they will be useful in continuum mechanics or pure geometry as well. The algebraic structure of the formulas is of some interest in its own right: the complicated dependence on the dimension m, the exotic coupling constant c, and the trace and transpose of the potential V are quite different from the second-order nonexotic case. The relative simplicity of (51) gives some hope of finding an insight into the nature of $a_n(x)$ for general n, generalizing Theorem 1. The most satisfactory outcome of a massive computer calculation often is a qualitative discovery that renders the calculation itself unnecessary in hindsight; that has not yet happened here, but one may hope, and keep looking.

Acknowledgments. The computations were done on a Sun 3/60 workstation with *Mathematica* software, both provided by a grant from Sun Microsystems, extended by Wolfram Research Inc. I am especially grateful to S. M. Christensen and L. Parker. Christensen posed this problem to me in 1980. He and Parker helped me obtain the computer, and they developed the software [29] that has proved to be necessary in solving this and similar problems. Parker provided advice and comfort during the calculation of (56). G. Kennedy collaborated on the key paper [16]. S. Mock and D. Potts wrote programs enumerating the permutations needed to construct $g_{2\beta}$. V. P. Gusynin proved in [23–24] that the intrinsic pseudodifferential calculus can indeed be used to get concrete results. He also caught an error in [21] before publication. I have learned an immense amount from P. B. Gilkey in public and private communications, and I am grateful to him and T. P. Branson for generously including me as a coauthor of [21]. I thank the organizers of this Colloquium for the opportunity to communicate this research.

Bulletin. Since the Colloquium, P. B. Gilkey (private communication) has reported a calculation of $a_2(H)$ for $k = 1$ by functorial methods, extending Theorems 1 and 2.

REFERENCES

1. I. G. Avramidi, A covariant technique for the calculation of the one-loop effective action, *Nucl. Phys. B* **355**, 712–754 (1991).
2. H. P. Baltes and E. R. Hilf, *Spectra of Finite Systems,* Bibliographisches Institut, Mannheim, 1976.
3. N. H. Barth and S. M. Christensen, Quantizing Fourth Order Gravity Theories: the Functional Integral, *Phys. Rev. D* **28**, 1876–1893 (1983).
4. A. O. Barvinsky and G. A. Vilkovisky, The generalized Schwinger–DeWitt technique in gauge theories and quantum gravity, *Phys. Reports* **119**, 1–74 (1985).
5. J. Bokobza-Haggiag, Opérateurs Pseudo-Différentiels sur une Variété Différentiable, *Ann. Inst. Fourier (Grenoble)* **19**, 125-177 (1969).
6. T. P. Branson, P. B. Gilkey, B. Ørsted, and A. Pierzchalski, Heat equation asymptotics of a generalized Ahlfors Laplacian on a manifold with boundary, to appear.
7. L. Brekhovskikh and V. Goncharov, *Mechanics of Continua and Wave Dynamics*, Springer, Berlin, 1985.
8. P. A. Carinhas and S. A. Fulling, Computational asymptotics of fourth-order operators, in *Asymptotic and Computational Analysis*, ed. by R. Wong, Marcel Dekker, New York, pp. 601–617.
9. S. M. Christensen, Vacuum Expectation Value of the Stress Tensor in an Arbitrary Curved Background: The Covariant Point-Separation Method, *Phys. Rev. D* **14**, 2490–2501 (1976).
10. S. M. Christensen, Regularization, Renormalization, and Covariant Geodesic Point-Separation, *Phys. Rev. D* **17**, 946–963 (1978).
11. L. Drager, *On the Intrinsic Symbol Calculus for Pseudo-Differential Operators on Manifolds*, Ph.D. Dissertation, Brandeis University, 1978.
12. S. A. Fulling, *Aspects of Quantum Field Theory in Curved Space-Time*, Cambridge University Press, Cambridge, 1989.
13. S. A. Fulling, The analytic approach to recursion relations, *J. Symbolic Computation* **9**, 73–85 (1990).
14. S. A. Fulling and G. Kennedy, A closed form for the symbol of the resolvent parametrix of an elliptic operator, in *Differential Equations and Mathematical Physics* (Lecture Notes in Math. No. 1285), ed. by I. W. Knowles and Y. Saito, Springer, Berlin, 1987, pp. 126–133.
15. S. A. Fulling and G. Kennedy, A closed form for the intrinsic symbol of the resolvent parametrix of an elliptic operator, in *The Physics of Phase Space* (Lecture Notes in Physics No. 278), ed. by Y. S. Kim and W. W. Zachary, Springer, Berlin, 1987, pp. 407–409.

16. S. A. Fulling and G. Kennedy, The resolvent parametrix of the general elliptic linear differential operator: A closed form for the intrinsic symbol, *Transac. Am. Math. Soc.* **310**, 583–617 (1988).

17. P. B. Gilkey, *The Index Theorem and the Heat Equation*, Publish or Perish, Boston, 1974.

18. P. B. Gilkey, The Spectral Geometry of a Riemannian Manifold, *J. Diff. Geom.* **10**, 601–618 (1975).

19. P. B. Gilkey, Recursion relations and the asymptotic behavior of the eigenvalues of the Laplacian, *Compos. Math.* **38**, 201–240 (1979).

20. P. B. Gilkey, *Invariance Theory, the Heat Equation, and the Atiyah-Singer Index Theorem*, Publish or Perish, Wilmington, 1984.

21. P. B. Gilkey, T. P. Branson, and S. A. Fulling, Heat equation asymptotics of "nonminimal" operators on differential forms, *J. Math. Phys.* **32**, 2089–2091 (1991).

22. P. Greiner, An Asymptotic Expansion for the Heat Equation, *Arch. Rat. Mech. Anal.* **41**, 163–218 (1971).

23. V. P. Gusynin, Seeley–Gilkey coefficients for the fourth-order operators on a Riemannian manifold, *Nucl. Phys. B* **333**, 296–316 (1990).

24. V. P. Gusynin, E. V. Gorbar, and V. V. Romankov, Heat kernel expansion for nonminimal differential operators and manifolds with torsion, *Nucl. Phys. B* **362**, 449–471 (1991).

25. V. D. Kupradze, *Potential Methods in the Theory of Elasticity*, Israel Program of Scientific Translation, Jerusalem, 1965.

26. H. W. Lee and P. Y. Pac, Higher-derivative operators and DeWitt's WKB ansatz, *Phys. Rev. D* **33**, 1012–1017 (1986).

27. H. P. McKean and I. M. Singer, Curvature and the eigenvalues of the Laplacian, *J. Diff. Geom.* **1**, 43–69 (1967).

28. S. Minakshisundaram and Å. Pleijel, Some Properties of the Eigenfunctions of the Laplace-Operator on Riemannian Manifolds, *Canad. J. Math.* **1**, 242–256 (1949).

29. L. Parker and S. M. Christensen, *MathTensor: A System for Doing Tensor Analysis by Computer*, MathSolutions, Chapel Hill, N.C., 1991.

30. L. Parker and D. J. Toms, Renormalization-group analysis of grand unified theories in curved spacetime, *Phys. Rev. D* **29**, 1584–1608 (1974).

31. R. T. Seeley, Complex Powers of an Elliptic Operator, in *Singular Integrals* (Proc. Symp. Pure Math. No. 10), American Mathematical Soc., Providence, 1967, pp. 288-307.

32. H. Weyl, Das asymptotische Verteilungsgesetz der Eigenschwingungen eines beliebig gestalteten elastischen Körpers, *Rend. Circ. Mat. Palermo* **39**, 1–49 (1915).

33. H. Weyl, Ramifications, old and new, of the eigenvalue problem, *Bull. Amer. Math. Soc.* **56**, 115–139 (1950).

34. H. Widom, A Complete Symbolic Calculus for Pseudodifferential Operators, *Bull. Sci. Math.* **104**, 19–63 (1980).

35. S. Wolfram, *Mathematica: A System for Doing Mathematics by Computer*, 2nd ed., Addison–Wesley, Redwood City, Calif., 1991.

Third International Colloquium on Differential Equations pp. 77-87 (1993)
D. Bainov and V. Covachev (Eds)

Self Adjoint Differential Equations and Karamata Functions

L. J. Grimm and L. M. Hall
University of Missouri-Rolla
Rolla, MO 65401, USA

ABSTRACT. A natural problem in the asymptotic theory of differential equations is the determination of the regularity of variation of solutions from the coefficients of the equation. Although some work has been done on special nonlinear equations, such as the Thomas-Fermi equation, considerably more results are available in the linear case. Recently V. Marić and M. Tomić have characterized the solutions of the equation $y'' - f(x)\,y = 0$ $(f \geq 0)$ according to the integrability properties of f. In this paper, we extend the results of Marić and Tomić to more general second-order linear equations and also functional differential equations. Our results are illustrated by several examples.

KEYWORDS: regular variation, Karamata function, self adjoint differential equation, functional-differential equation.

INTRODUCTION

The subject of regular variation has been developed extensively since it was introduced in the early 1930's by J. Karamata in his work on Tauberian theorems. The Karamata theory, which includes the related concepts of slowly and rapidly varying functions, has proved to be effective in treating a variety of problems in analysis, including the theory of entire functions, probability theory, and Fourier analysis. See, for example, the monographs [1], [2], [7].

A natural problem in the asymptotic theory of differential equations is the determination of the regularity of variation of solutions from the coefficients of the equation. Although some work has been done on special nonlinear equations, such as the Thomas-Fermi equation [3], considerably more results are available in the linear case; see Omey [6], Marić and Tomić [4]. Recently Marić and Tomić [5] have characterized the solutions of the equation $y'' - f(x)\,y = 0$ $(f \geq 0)$ according to the integrability properties of f. In this note, we extend the results of Marić and Tomić to more general second-order linear equations.

We begin with definitions of Karamata functions.

Definition. Let g(x) be a positive measurable function on $[x_0, \infty)$. If, for each $\lambda > 1$, there exists α, $-\infty \leq \alpha \leq \infty$, such that

$$\lim_{x\to\infty} \frac{g(\lambda x)}{g(x)} = \lambda^{\alpha}$$

then, at infinity, g(x) has index α and is said to be

i) slowly varying if $\alpha = 0$,

ii) regularly varying if $\alpha \neq 0$, $-\infty < \alpha < \infty$,

iii) rapidly varying if $\alpha = -\infty$ or $+\infty$.

We shall use the following properties of these functions (see [2], [5], or [7]) in the rest of this paper.

Lemma 1. (Representation Theorem) The function L(x) is slowly varying if and only if it can be represented in the form

$$L(x) = c(x)\exp\left\{-\int_a^x \frac{\varepsilon(t)}{t}dt\right\}$$

for some a > 0, where c and ε are measurable, $\lim_{x\to\infty} c(x) = c$ *and* $\lim_{x\to\infty} \varepsilon(x) = 0$.

If $c(x) \equiv c$, then L(x) is called normalized slowly varying.

Notation. Slowly varying functions (svf) will be denoted by L(x), and normalized slowly varying functions (nsvf) will be denoted by $L_0(x)$.

Lemma 2. $L_0(x)$ is an nsvf if and only if

$$\lim_{x\to\infty} \frac{xL_0'(x)}{L_0(x)} = 0.$$

Lemma 3. Let L(x) be an svf. Then, for each $\eta > 0$,

$$\lim_{x\to\infty} x^{\eta} L(x) = \infty \quad and \quad \lim_{x\to\infty} x^{-\eta} L(x) = 0.$$

Lemma 4. Let L(x) be an svf. If L is differentiable and convex, that is, if L'(x) is increasing, then

$$\lim_{x\to\infty} \frac{xL'(x)}{L(x)} = 0.$$

Lemma 5. Suppose that $\quad f(x) = \int_x^\infty \psi(t)dt \equiv x^{-\alpha}L(x),$

where ψ is nonincreasing, $\alpha > 0$, and L is an svf. Then $\lim_{x\to\infty} \frac{-xf'(x)}{f(x)} = \alpha$.

Lemma 6. Suppose $g(x) > 0$, $g'(x)$ is increasing and $\lim_{x\to\infty} g(x) = 0$.

Then g is rapidly varying if and only if $\lim_{x\to\infty} \frac{-xg'(x)}{g(x)} = \infty$.

QUALITATIVE BEHAVIOR OF SOLUTIONS

We first consider the equation

$$(p(x)y')' + f(x)y = 0 \tag{2.0}$$

where f is continuous, and p is positive and differentiable on $[x_0, \infty)$. If (2.0) is nonoscillatory at $x = \infty$, then a nonzero solution y(x) for which

$$\int^{\infty} \frac{dx}{p(x)y^2(x)} = \infty$$

is called a principal solution of (2.0), and a solution independent of y(x) is called a nonprincipal solution. For instance, e^{-2t} and e^{2t} are principal and nonprincipal solutions for the equation $y'' - y = 0$. The above integral will converge with a nonprincipal solution in place of y(x). If, in addition, (2.0) is disconjugate on $[x_0, \infty)$, and $y_0(x)$ is a principal solution, then y_0 (or $-y_0$) will satisfy $y_0(x) > 0$, $y_0'(x) \leq 0$ on $[x_0, \infty)$. We show how a result of Marić and Tomić [5] can extend to characterize the asymptotic behavior of the principal solution in the case where $f(x) \leq 0$. In this case, Sturm's theorem ensures the disconjugacy of (2.0), and so the existence of a positive decreasing solution is guaranteed. Hence, for the remainder of this paper, we consider the equation

$$\left(p(x)y'\right)' - f(x)y = 0 \tag{2.1}$$

where f is positive and continuous, and p is positive and differentiable on $[x_0, \infty)$.

Theorem 2.1 Let y be a positive decreasing solution of (2.1) on $[x_0, \infty)$, and let p be nondecreasing and bounded on $[x_0, \infty)$. Then y is a slowly varying function if and only if

$$\lim_{x\to\infty} x\int_x^{\infty} f(t)dt = 0. \tag{2.2}$$

Proof: Suppose that y is slowly varying. Then, since y is convex, Lemmas 2 and 4 imply that y is an nsvf and so

$$\frac{xy'}{y} \to 0 \quad as \quad x \to \infty.$$

Equation (2.1) can be written in the form $\dfrac{p'y'}{y} + \dfrac{py''}{y} = f$,

or
$$\frac{p'y'}{y} + p\left(\frac{y'}{y}\right)' + p\left(\frac{y'}{y}\right)^2 = f.$$

Integrate and multiply by x to obtain

$$x\int_x^\infty \frac{p'y'}{y}dt + x\int_x^\infty p\left(\frac{y'}{y}\right)' dt + x\int_x^\infty \frac{p}{t^2}\left(t\frac{y'}{y}\right)^2 dt = x\int_x^\infty f\,dt > 0.$$

Since y is slowly varying and p is bounded, the second and third integrals on the left tend to zero as $x \to \infty$; the first is nonpositive. Hence

$$x\int_x^\infty f\,dt \to 0 \quad as \quad x \to \infty.$$

Conversely, suppose

$$x\int_x^\infty f\,dt \to 0 \quad as \quad x \to \infty.$$

Then integration of (2.1) from x to ∞ yields $py'\big|_x^\infty = \int_x^\infty f\,y\,dt$.

Since y is decreasing, and $y' \to 0$ as $x \to \infty$, we have

$$-xp(x)\frac{y'(x)}{y(x)} = \frac{x}{y(x)}\int_x^\infty f(t)y(t)dt \le x\int_x^\infty f(t)\,dt$$

and so there is a $\xi > x$ such that $-xp(x)\dfrac{y'(x)}{y(x)} = x\int_x^\xi f(t)dt \equiv \varepsilon(x)$

which tends to zero as $x \to \infty$. Thus $y(x) = k\exp\left(-\int_a^x \frac{\varepsilon(t)}{t\,p(t)}dt\right)$

and by Lemma 1, y is an nsvf. This completes the proof of Theorem 2.1.

Theorem 2.2 Let y be a solution of (2.1) on $[x_0,\infty)$ such that y is positive and decreasing, and let p be a positive nondecreasing nsvf on $[x_0,\infty)$, with $\lim_{x\to\infty} p(x) = 1$. Then y will be a regularly varying function of index $\alpha = (1 - \sqrt{(1+4c)})/2$ if and only if

$$\lim_{x\to\infty} x\int_x^\infty f(t)\,dt = c. \tag{2.3}$$

Remark. The restriction $\lim_{x\to\infty} p(x) = 1$ is a convenient normalization and can be replaced by the requirement that p be bounded.

Proof : As in the proof of Theorem 2.1, integration of (2.1) leads to

$$x\int_x^\infty \frac{p'y'}{y}dt + x\int_x^\infty p\left(\frac{y'}{y}\right)' dt + x\int_x^\infty \frac{p}{t^2}\left(t\frac{y'}{y}\right)^2 dt = x\int_x^\infty f\,dt. \tag{2.4}$$

Suppose y is regularly varying of index α. Then $y = x^\alpha L(x)$, where $\alpha < 0$ (by Lemma 3) and L is an svf. Using Lemma 5, we can show that $xL'/L \to \infty$ as $x \to \infty$, so that L is an nsvf. Integrating the second integral in (2.4) by parts and simplifying gives

$$-xp\frac{y'}{y} + x\int_x^\infty \frac{p}{t^2}\left(t\frac{y'}{y}\right)^2 dt = x\int_x^\infty f(t)\ dt\,,$$

and since $y'/y = L'/L + \alpha/x$ and $p(x) \to 1$ as $x \to \infty$, we get $x\int_x^\infty f(t)dt \to \alpha^2 - \alpha$ as $x \to \infty$. Now assume

$$x\int_x^\infty f(t)\ dt \to c \quad as \quad x \to \infty$$

and let α be the negative root of $\alpha^2 - \alpha = c$. For $k > 0$, let

$$y(x) = k\exp\left(-\int_a^x \frac{\eta(t)-\alpha}{t}\,dt\right)$$

so that

$$\eta(x) \equiv \frac{-xy'}{y} + \alpha$$

and η satisfies

$$y(x)\left[\left(\frac{\eta(x)-\alpha}{-x}\right)p'(x) + p(x)\left(\frac{\eta(x)-\alpha}{-x}\right)^2 + p(x)\left(\frac{\eta(x)-\alpha-x\eta'(x)}{x^2}\right)\right] = f(x)y(x) \tag{2.5}$$

Now integrating (2.1) and using the definition of η leads to

$$0 < (\eta(x)-\alpha)p(x) \le x\int_x^\infty f(t)dt \to \alpha^2 - \alpha$$

and so η is bounded. If we now divide (2.5) by py, integrate, and multiply by x, we get

$$x\int_x^\infty \frac{p'}{p}\left(\frac{\eta(t)-\alpha}{-t}\right)dt + x\int_x^\infty\left(\frac{\eta(t)-\alpha}{-t}\right)^2 dt + x\int_x^\infty\left(\frac{\eta(t)-\alpha-t\eta'(t)}{t^2}\right)dt = x\int_x^\infty \frac{f(t)}{p(t)}\,dt \tag{2.6}$$

where the right side can be expressed as $c + \delta(x)$, $\delta(x) \to 0$ as $x \to \infty$. Integrating the term containing η' by parts, and using the fact that $\alpha^2 - \alpha = c$, (2.6) becomes

$$x\int_x^\infty \frac{p'(t)}{p(t)}\left(\frac{\eta(t)-\alpha}{-t}\right)dt + \eta(x) + x\int_x^\infty \frac{\eta(t)(\eta(t)-2\alpha)}{t^2}\,dt = \delta(x). \tag{2.7}$$

It is easy to show that the first integral tends to 0 as x tends to ∞, and so it can be absorbed into the $\delta(x)$ term. Thus the other integral converges. Since we know that $\eta(t) - \alpha > 0$, $\eta(t) - 2\alpha > 0$ as well, and so the sign of the last two terms on the left side of (2.7) depends only on the sign of η. Clearly, if η is ultimately of constant sign, $\eta(x) \to 0$ as $x \to 0$.

Suppose $\eta(x)$ has infinitely many zeros x_i^0. Let x_i be the nearest point where η has a local maximum to the left of x_i^0. From (2.5) we have, after some calculations,

$$x_i\int_{x_i}^{x_i^0}\frac{p'}{p}\left(\frac{\eta(t)-\alpha}{-t}\right)dt+x_i\int_{x_i}^{x_i^0}\left(\frac{\eta(t)-\alpha}{-t}\right)^2 dt-x_i\int_{x_i}^{x_i^0}\frac{\alpha}{t^2}\,dt+\eta(x_i)=x_i\int_{x_i}^{x_i^0}\frac{f(t)}{p(t)}\,dt$$

As before, for x_i large, the first integral tends to zero and the right side tends to c, so

$$x_i\int_{x_i}^{x_i^0}\frac{\eta^2(t)-2\alpha\eta(t)+\alpha^2-\alpha}{t^2}\,dt+\eta(x_i)=c\left(\frac{1}{x_i}-\frac{1}{x_i^0}\right)x_i-\gamma(x_i^0)\left(\frac{x_i}{x_i^0}\right)+\gamma(x_i)$$

where $\gamma(x)\to 0$ as $x\to 0$. Further simplification and the use of the intermediate value theorem yields

$$\eta(\xi)(\eta(\xi)-2\alpha)\left(1-\frac{x_i}{x_i^0}\right)+\eta(x_i)=\gamma(x_i)-\gamma(x_i^0)\left(\frac{x_i}{x_i^0}\right)$$

The right side $\to 0$ as $x_i\to\infty$, and the left side is the sum of positive terms, so $\eta(x_i)\to 0$ as $x_i\to\infty$. This, along with Lemma 1, means that $y(x)=x^{\alpha}L_0(x)$, and so y is regularly varying with index α. This completes the proof of Theorem 2.2.

Theorem 2.3. Let y be a solution of (2.1) on $[x_0,\infty)$ such that y is positive and decreasing, and let p be a positive nondecreasing nsvf on $[x_0,\infty)$, with $\lim_{x\to\infty}p(x)=1$. Then y will be rapidly varying if and only if for each $k>1$,

$$\lim_{x\to\infty} x\int_x^{kx} f(t)\,dt=\infty \tag{2.8}$$

Proof: Suppose that y is rapidly varying, and let $k>1$ be given. Since y is decreasing, integration of (2.1) from x to kx yields:

$$\left(\frac{-xy'(x)}{y(x)}\right)p(x)\left(1-\left(\frac{p(kx)}{p(x)}\right)\left(\frac{y'(kx)}{y'(x)}\right)\right)\le x\int_x^{kx} f(t)\,dt$$

Since y is rapidly varying, the first term in parentheses $\to\infty$ by Lemma 6, and so the right side $\to\infty$ provided that $y'(kx)/y'(x)$ does not $\to 1$ as $x\to\infty$. This is proved in Marić and Tomić [5].

Now suppose that $\{x_i\}$ is an arbitrary sequence on which (2.8) holds and such that $x_i\to\infty$ as $i\to\infty$. For y a solution of (2.1) with y positive and decreasing, define

$$\phi(x)\equiv\frac{-x\,(p_o y)'(x)}{(p_o y)(x)}$$

It can be shown that $\liminf_{i\to\infty}\phi(x_i)>0$ and so we only need to consider the cases where the lim sup is infinite and finite, respectively.

Suppose $\limsup_{i\to\infty}\phi(x_i) = \infty$. Then for $\delta > 0$, there is a ξ_i, $x_i \le \xi_i \le (1+\delta)x_i$, such that

$$\frac{y(x_i)-y(x_i+\delta x_i)}{y(x_i+\delta x_i)} = \frac{-\delta x_i y'(\xi_i)}{y(x_i+\delta x_i)} \ge \frac{-\delta x_i y'(x_i+\delta x_i)}{y(x_i+\delta x_i)}$$

and so by our assumption,

$$\frac{y(x_i)}{y(x_i+\delta x_i)} - 1 \to \infty \; as \; x_i \to \infty.$$

Thus, on $\{x_i\}$, $y(x_i)/y(\lambda x_i) \to \infty$, where $\lambda = 1 + \delta$.

Now suppose $\limsup_{i\to\infty}\phi(x_i) = M < \infty$. For $k > 1$, integration of (2.1) from x to kx gives

$$\frac{x\, y'(kx)p(kx)}{y(kx)p(x)} \ge \frac{x\, y'(x)}{y(kx)} + \frac{x}{p(x)}\int_x^{kx} f(t)\,dt$$

The left side of this inequality is < 0, and the term containing the integral $\to \infty$, so the remaining term must $\to -\infty$ as $x \to \infty$. Thus on $\{x_i\}$, we have

$$\left(\frac{-x_i y'(x_i)}{y(x_i)}\right)\left(\frac{y(x_i)}{y(kx_i)}\right) \to \infty \quad as \quad i \to \infty.$$

By our assumption, the first factor in the limit is bounded, and so the second must approach ∞.

We have proved that, for any sequence $\{x_i\}$ on which (2.8) holds, we have $y(x_i)/y(x+\delta x_i) - 1 \to \infty$ as $x \to \infty$. Thus y is rapidly varying, and the proof is complete.

EQUIVALENCE OF THE INTEGRAL CONDITIONS

It would be convenient if the limits on the integral in Theorem 2.3 were the same as those in Theorems 2.1 and 2.2, but this cannot be the case. We will construct a positive continuous function f such that

$$\lim_{x\to\infty} x \int_x^{\infty} f(t)\,dt = \infty, \text{ but } \lim_{x\to\infty} x \int_x^{kx} f(t)\,dt \ne \infty$$

for some $k > 1$. The following theorem provides some insight into how the integral conditions are related.

Theorem 2.4. If, for $k > 1$,

(1) $\limsup_{x\to\infty} x \int_x^{kx} f(t)\,dt \le L^+$, *then* $\limsup_{x\to\infty} x \int_x^{\infty} f(t)\,dt \le \dfrac{k L^+}{k-1}$,

(2) $\liminf_{x\to\infty} x \int_x^{kx} f(t)\,dt \geq L^-$, *then* $\liminf_{x\to\infty} x \int_x^{\infty} f(t)\,dt \geq \dfrac{kL^-}{k-1}$,

(3) $\lim_{x\to\infty} x \int_x^{kx} f(t)\,dt = L(k)$, *then* $\lim_{x\to\infty} x \int_x^{\infty} f(t)\,dt = L(k)\dfrac{k}{k-1}$ *and there exists a number c such that* $L(k) = c\left(\dfrac{k-1}{k}\right)$.

Proof: For $\varepsilon > 0$ and x large,

$$x \int_{k^n x}^{k^{n+1}x} f(t)\,dt \leq \frac{L^+ + \varepsilon}{k^n}$$

so we have

$$x \int_x^{\infty} f(t)\,dt = \sum_0^{\infty} \int_{k^n x}^{k^{n+1}x} f(t)dt \;\leq\; \sum_{n=0}^{\infty} \frac{L^+ + \varepsilon}{k^n} = \frac{k(L^+ + \varepsilon)}{k-1}.$$

This proves (1) and, with obvious modifications, (2). (1) and (2) give (3).

Thus, if $\lim_{x\to\infty} x \int_x^{\infty} f(t)\,dt = \infty$, then $\limsup_{x\to\infty} x \int_x^{\infty} f(t)\,dt = \infty$ also, and so we need a function for which the first limit is infinite, but the second (with "lim sup" replaced by "lim") does not exist as an extended real number. Consider the following piecewise linear, continuous, nonnegative function:

$$g(x) = \begin{cases} g_n(x) & \text{on the intervals } \; 2^n \leq x < 2^{n+1},\ n \geq 0 \text{ and even,} \\ 0 & \text{on the intervals } \; 2^n \leq x < 2^{n+1},\ n \geq 0 \text{ and odd,} \end{cases}$$

where for $2^n \leq x < 2^{n+1}$, $n \geq 0$ and even, the graph of $g_n(x)$ consists of the line segments from $(2^n, 0)$ to $(2^{n+1/2}, 2^{-3n/2})$ to $(2^{n+1}, 0)$. Then if 2^{2n} is the first even power of 2 to the right of x, we get

$$\int_x^{\infty} g(x) = \sum_{k=n}^{\infty} 2^{-k-1} = \frac{1}{2^n}, \textit{ and so } x\int_x^{\infty} g(x) \geq 2^{n-2}, \textit{ which} \to \infty \textit{ as } x \to \infty.$$

However, on the sequence $\{2^{2k+1}\}$ it is clear that $x \int_x^{2x} g(t)\,dt = 0$, and thus it is not true that for all $k > 1$, $\lim_{x\to\infty} x \int_x^{kx} g(t)\,dt = \infty$. The function g is not positive, but the example can be modified by simply adding x^{-2} without losing any vital features.

While the above example shows that the integral condition in Theorem 2.3 cannot be changed to match the conditions in Theorems 2.1 and 2.2, we can use Theorem 2.4 to show that the conditions in Theorems 2.1 and 2.2 can be replaced by the condition in Theorem 2.3.

The next result follows immediately from Theorem 2.4.

Corollary 2.5. $\lim_{x\to\infty} x \int_x^\infty f(t)\,dt = 0$ if and only if $\lim_{x\to\infty} x \int_x^{kx} f(t)\,dt = 0$, and

$\lim_{x\to\infty} x \int_x^\infty f(t)\,dt = c$ if and only if $\lim_{x\to\infty} x \int_x^{kx} f(t)\,dt = c(k-1)/k$.

Thus, Theorems 2.1 and 2.2 can be stated in terms of the integral from x to kx, but the original integral conditions are easier to verify, so we will not change the statements of those theorems.

EXAMPLES

Example 1. Let $p(x) = x/(x+1)$, and $f(x) = \dfrac{x\log(x) + 2x + 2}{x(x+1)^2 \log^2(x)}$
Here

$$\lim_{x\to\infty} x \int_x^\infty f(t)\,dt = 0,$$

and so by Theorem 2.1, all positive decreasing solutions are slowly varying. It is easy to verify that $y(x) = (\log(x))^{-1}$ is one such solution.

Example 2. Let p be as in Example 1, and let $f(x) = \dfrac{6x+4}{x(x+1)^2}$.
Now

$$\lim_{x\to\infty} x \int_x^\infty f(t)\,dt = 6,$$

and so by Theorem 2.2, y is regularly varying of index -2. The reader can verify that $y(x) = x^{-2}$ is a solution.

Example 3. Consider Airy's equation, $y'' - x\,y = 0$. The integral condition in Theorem 2.3 is clearly satisfied, and so positive decreasing solutions are rapidly varying. Thus, the Airy function Ai(x) is rapidly varying as $x \to \infty$.

LINEAR FUNCTIONAL DIFFERENTIAL EQUATIONS

Some related results can be obtained for functional-differential equations. For example, consider the equation

$$y''(x) = f(x)y(x) + g(x)y(h(x)) \tag{5.1}$$

where f and g are nonnegative and f, g, and h are continuous on (x_0, ∞). The following result is similar to Theorem 2.1.

Theorem 5.1. Let $h(x) \geq x$, and let y be a positive decreasing solution of (5.1) on (x_0,∞). Then y is a svf if and only if $\lim_{x\to\infty} x \int_x^\infty f(t)\,dt = \lim_{x\to\infty} x \int_x^\infty g(t)\,dt = 0$.

Proof. Suppose that y is slowly varying. Since y is convex, y is an nsvf and hence $xy'/y \to 0$ as $x \to \infty$. Write (5.1) in the form $y''(x)/y(x) = f(x) + g(x)y(h(x))/y(x)$, integrate and multiply by x to obtain

$$x\int_x^\infty \left(\frac{y'(t)}{y(t)}\right)' dt + x\int_x^\infty \frac{1}{t^2}\left(\frac{ty'(t)}{y(t)}\right)^2 dt = x\int_x^\infty f(t)\,dt + x\int_x^\infty g(t)\frac{y(h(t))}{y(t)}\,dt.$$

Since y is slowly varying, the integrals on the left $\to$ zero as $x \to \infty$. Since both of the integrals on the right are nonnegative, each of these must also $\to 0$.

Conversely, suppose that both $x\int_x^\infty f(t)\,dt$ and $x\int_x^\infty g(t)\,dt \to 0$ as $x \to \infty$.

Integration of (5.1) from x to ∞ yields $y'(\infty) - y'(x) = \int_x^\infty f(t)y(t)\,dt + \int_x^\infty g(t)y(h(t))\,dt$ Since $h(t) \geq t$ and y is decreasing with $y' \to 0$ as $x \to \infty$, we have

$$-\frac{xy'(x)}{y(x)} \leq x\int_x^\infty f(t)\,dt + x\int_x^\infty g(t)\,dt = x\int_x^\infty (f(t)+g(t))\,dt \equiv \varepsilon(x),$$

which $\to 0$ as $x \to \infty$.

The proof of the first part of Theorem 5.1 also yields the following result.

Theorem 5.2. Let $h(x) > x_0$ for $x \in (x_0, \infty)$. If y is a slowly varying positive decreasing solution of (5.1) on (x_0, ∞), then $\lim_{x\to\infty} x\int_x^\infty f(t)\,dt = \lim_{x\to\infty} x\int_x^\infty g(t)\,dt = 0$.

The following example shows that the converse of Theorem 5.2 is false for the retarded argument case.

Example 4. Consider the equation $y''(x) = \left[\frac{1}{x^2} + 1 + 2x^2\right] e^{\frac{-3x^2}{4}} y\left(\frac{x}{2}\right)$

for $0 < x < \infty$. It is easy to see that

$$\lim_{x\to\infty} x\int_x^\infty \left[\frac{1}{t^2} + 1 + 2t^2\right] e^{\frac{-3t^2}{4}}\,dt = 0$$

while the positive decreasing function $\phi(x) = (1/x)\exp(-x^2)$ is a rapidly varying solution.

The authors wish to thank R. J. Sacker and S. Y. Trimble for stimulating discussions while this paper was being prepared.

REFERENCES

1. N. H. Bingham, C. M. Goldie, J. L. Teugels, Regular Variation, Encyclopedia of Mathematics and its Applications, vol. 27, Cambridge University Press, Cambridge, 1987.

2. J. L. Geluk, L. de Haan, Regular Variation, Extensions and Tauberian Theorems, CWI Tract 40, Stichting Mathematisch Centrum, Centrum voor Wiskunde en Informatica, Amsterdam, 1987.

3. V. Marić and M. Tomić, Asymptotics of solutions of a generalised Thomas-Fermi equation, J. Differential Equations, 35 (1980), 36-44.

4. V. Marić and M. Tomić, A trichotomy of solutions of second order linear differential equations, Rev. Res. Fac. Sci. Univ. Novi Sad Math. Ser. 14 (1984), 1-11.

5. V. Marić and M. Tomić, A classification of solutions of second order linear differential equations by means of regularly varying functions, Publ. Inst. Math. (Beograd) (N. S.) 48 (62) (1990), 199–207.

6. E. Omey, Regular variation and its applications to second order linear differential equations, Bull. Soc. Math. Belg. , Ser. B, 33 (1981), 207-229

7. E. Seneta, Regularly Varying Functions, Lecture Notes in Mathematics, vol. 508, Springer-Verlag, Berlin, 1976.

Third International Colloquium on Differential Equations pp. 89-93 (1993)
D. Bainov and V. Covachev (Eds)

ORDINARY DIFFERENTIAL EQUATIONS IN FRECHET SPACES

Gerd Herzog and Roland Lemmert
Mathematisches Institut I der Universität Karlsruhe,
Englerstraße 2, D-7500 Karlsruhe

Abstract: After a short survey on existence and non-existence results for the Cauchy problem

$$(\star) \qquad \begin{cases} u' = f(t,u), \\ u(0) = x_0 \end{cases}$$

in Banach spaces of infinite dimension, we discuss the difficulties that arise when studying $(\star)$ in Fréchet spaces E. Here, Lipschitz conditions on f (with respect to any metric defining the topology of E) are not sufficient to imply existence of solutions in general, even linear equations need not be solvable or uniquely solvable.

We give conditions on f in terms of seminorms that ensure existence, uniqueness, convergence of successive approximations and continuous dependence. The main tool is the theory of row-finite matrices as developed by Ulm.

Keywords: Cauchy problem, Peano's theorem

Let E be a (real or complex) locally convex space and $f : [0,T] \times E \to E$ be continuous, $x_0 \in E$ fixed. We are looking for conditions on f and E such that

$$(\mathrm{P}) \qquad \begin{cases} u'(t) = f(t,u(t)), \\ u(0) = x_0 \end{cases}$$

is solvable, locally solvable and uniquely solvable, respectively.

Combining Peano's existence theorem and a theorem of Godunov [4], we get the following

Theorem. *For a Banach space E, the following assertions are equivalent:*

a) $\dim E < \infty$,

b) (P) *is locally solvable for each continuous f,*

c) (P) *is solvable for each bounded continuous f.*

So in each case, if $\dim E = \infty$, additional hypotheses on f and E are necessary in order b) or c) of the above theorem hold.

For Banach spaces E, an abundance of such conditions has been given in recent years; we only list a (of course incomplete) number of names: Ambrosetti, Cellina, Deimling, Pavel, Schechter, Schmidt, Steçenko, Szufla, Dubinsky, Godunov, Hu, Lemmert, Li, Volkmann, and Walter. For further information, the reader should consult the books by Deimling [2], Martin [8] or Pavel [10]. Such conditions are given in terms of measures of noncompactness, (one-sided) Lipschitz conditions and monotonicity properties of f in ordered Banach spaces. In passing we mention that the theorem of Picard-Lindelöf holds without change in each Banach space.

Now, if we consider more general locally convex topological vector spaces, the situation is completely different. Here, even in the class of Montel-Fréchet spaces, b) and c) are no longer equivalent.

Example 1. Let $E = \mathbb{R}^{\mathbb{N}} = \{x = (x_n)_{n\in\mathbb{N}},\, x_n \in \mathbb{R}\}$ be equipped with the topology of coordinatewise convergence. If we define $f : E \to E$ by $f_n(x) = x_n^2 + n$ and if (P) was locally solvable, we should have

$$u_n(t) = \frac{\sqrt{n}\cdot x_{0,n} + n\cdot \tan\left(\sqrt{n}\cdot t\right)}{\sqrt{n} - x_{0,n}\cdot \tan\left(\sqrt{n}\cdot t\right)}, \quad n \in \mathbb{N}, \quad t \in [0,\tau),$$

which is not possible.

On the other hand, Millionščikov [9] and later Dubinsky [3], following the ideas of Tychonoff [11, 12], proved:

Theorem. c) *is valid in each Montel space.*

Concerning non-Montel Fréchet spaces, one has only partial results. Following ideas of Cellina [1], Herzog [5] proved the following theorems.

Theorem. *If E is a non-reflexive Fréchet space,* c) *is not valid.*

Theorem. *If E has an infinite-dimensional closed normable subspace,* c) *is not valid.*

Unfortunately, there are reflexive Fréchet spaces [5, Beispiel 4.2] with no normable infinite-dimensional subspaces, so it is not clear if c) is not valid for all such spaces.

Let us now turn to the more delicate question of the validity of b); here, even in case E is a Montel-Fréchet space, an answer to this question seems to be unknown (in view of Example 1, it should perhaps be no).

As b) contains linear equations, too, let us quote the following examples.

Example 2. Let $E = \{x \in C^\infty([-1,1],\mathbb{R}),\ x^{(n)}(0) = 0,\ n \geq 0\}$ with the topology induced by the system $\{p_n\}$ of seminorms given by $p_n(x) = \max_{-1\leq s\leq 1} |x^{(n)}(s)|$. If we define $A : E \to E$ by $(A(x))(s) = x'(s)$, A is continuous, and (P) with $f(t,x) = Ax$ takes the form

$$u_t(s,t) = u_s(s,t), \quad -1 \leq s \leq 1, \quad 0 \leq t \leq t_0,$$

with $u(s,0) = x_0(s)$ a given function. Since the characteristics of the above partial differential equation are of the form $t = -s + \text{const.}$, we get $x_0(s) = 0$ for $0 \leq s \leq t_0$, a contradiction for an appropriate choice of x_0.

If we choose the usual Fréchet metric in the space E of Example 2, we easily deduce $d(Ax, Ay) \leq 2d(x,y)$ so that the right hand side of the differential equation is uniformly Lipschitz continuous, but (P) in general not solvable, as one might have expected.

Example 3. We set $E = C^\infty(\mathbb{R}, \mathbb{R})$ with the topology induced by the seminorms $p_n(x) = \max\{|x^{(j)}(s)| : 0 \leq j \leq n,\ |s| \leq n\}$, $n \in \mathbb{N}$, and A as in Example 2. Here, the corresponding initial value problem is uniquely solvable for each $x_0 \in E$, and is in fact given by $u(s,t) = x_0(s+t)$. This solution is, in general, not of the form $\sum_{n=0}^{\infty} \frac{A^n x_0}{n!} t^n$, as is seen by choosing x_0 such that $x_0^{(n)}(0) = n^n$.

Example 4. Let $E = \mathbb{R}^{\mathbb{N}}$ and A be the left shift, i.e., $(Ax)_n = x_{n+1}$. Each $x_0 \in C^\infty(\mathbb{R}_+)$ with $x_0^{(n)} = 0$, $n \geq 0$, produces a solution $u = (x_0, x_0', x_0'', \ldots)$ to $u' = Au$, $x(0) = 0$.

For later purposes, let us consider $E = \mathbb{C}^{\mathbb{N}}$ in more detail, cf. [6]. Here, a linear continuous operator A (we write $A \in L(E)$) can be thought of as a row-finite matrix $(a_{ij})_{i,j=1}^{\infty}$, where in each row an at most finite number of elements is different from zero. The topological dual can be indentified with $\mathbb{C}_{\mathbb{N}} = \{y = (y_n) : y_n \in \mathbb{C},\ y_n \neq 0$ for an at most finite number of n's$\}$. If $\mathbb{C}_{\mathbb{N}}$ is equipped with the weak* topology $\sigma(\mathbb{C}_{\mathbb{N}}, \mathbb{C}^{\mathbb{N}})$, we have (cf. [6])

Theorem. *Let $B \in L(\mathbb{C}_{\mathbb{N}})$ and $y_0 \in \mathbb{C}_{\mathbb{N}}$. Then the following assertions are equivalent:*

i) *$u' = Bu$, $u(0) = y_0$ has a solution.*

ii) *Here exists a polynomial q such that $q(B)y_0 = 0$.*

iii) *$\sum_{n=0}^{\infty} B^n y_0 \cdot \frac{t^n}{n!}$ is convergent in $\mathbb{C}_{\mathbb{N}}$.*

iv) *$u' = Bu$, $u(0) = y_0$ has a unique solution.*

So, $\mathbb{C}_{\mathbb{N}}$ is an example of a locally convex space, in which linear initial value problems are uniquely solvable if they are solvable at all. For B the right shift, the initial value problem is unsolvable for each $y_0 \neq 0$.

Property ii) of the above theorem gives rise to the following

Definition. The operator $B \in L(\mathbb{C}_{\mathbb{N}})$ is called locally algebraic, if ii) is valid for each $y_0 \in \mathbb{C}_{\mathbb{N}}$.

With this we have [6]:

Theorem. *For $A \in L(\mathbb{C}^{\mathbb{N}})$, the following statements are equivalent:*

i) *The transpose ${}^{T}\!A \in L(\mathbb{C}_{\mathbb{N}})$ is locally algebraic.*

ii) *$u' = Au$, $x(0) = x_0$ is uniquely solvable for each $x_0 \in \mathbb{C}^{\mathbb{N}}$.*

iii) *The spectrum $\sigma(A)$ is at most countable.*

iv) *$\sum_{n=0}^{\infty} A^n x_0 \frac{t^n}{n!}$ is convergent in $\mathbb{C}^{\mathbb{N}}$ for each $x_0 \in \mathbb{C}^{\mathbb{N}}$, $t \in \mathbb{R}$.*

Let us remark that in $\mathbb{C}^{\mathbb{N}}$ linear initial value problems are always solvable, cf. [5].

We turn again to the case where E is a Fréchet space with a system $(p_n)_{n=1}^{\infty}$ of seminorms. For each $x \in E$ we define $||x|| := (p_n(x))_{n=1}^{\infty} \in \mathbb{R}^{\mathbb{N}}$. $||\cdot||$ has the usual properties of a norm, if we interprete inequalities componentwise. A linear operator L is continuous if and only if there exists a real row-finite matrix A such that

$$||Lx|| \le A \cdot ||x||, \quad x \in E.$$

We are now able to prove the following analogue of Picard-Lindelöf's theorem.

Theorem. *Let $f : [0,T] \times E \to E$ be continuous, and let there exist $A \in L(\mathbb{R}^{\mathbb{N}})$ with an at most countable spectrum such that*

$$||f(t,x) - f(t,\bar{x})|| \le A||x - \bar{x}||, \quad x, \bar{x} \in E, \quad t \in [0,T].$$

Then (P) *has exactly one solution u, and the method of successive approximations converges.*

If (u_n) is the corresponding sequence of approximations, we have

$$||u_{n+1}(t) - u_n(t)|| \le L^n \frac{t^n}{n!}\, ||u_1(t) - u_0(t)||$$

and

$$||u(t) - u_n(t)|| \le \sum_{k=n}^{\infty} L^k \frac{t^k}{k!}\, ||u_1(t) - u_0(t)||,$$

$n \in \mathbb{N}$, $t \in [0,T]$.

References

1. A. Cellina, *On the nonexistence of solutions of differential equations in nonreflexive spaces*. Bull. Amer. Math. Soc. **78**, 1069–1072 (1972).

2. K. Deimling, *Ordinary differential equations in Banach spaces*. Lecture Notes in Mathematics, Vol. 596, Springer, Berlin - Heidelberg - New York (1977).

3. E. Dubinsky, *Differential equations and differential calculus in Montel spaces*. TAMS **110**, 1-21 (1964).

4. A.N. Godunov, *O teoreme Peano v banahovyh prstranstvah*. Funktional'nyi Analiz i Ego Prilozheniya **9**, 59–60 (1975).

5. G. Herzog, *Über gewöhnliche Differentialgleichungen in Fréchetráumen*. Dissertation, Karlsruhe (1992).

6. R. Lemmert and Ä. Weckbach, *Charakterisierungen zeilenendlicher Matrizen mit abzählbarem Spektrum*. Math. Z. **188**, 119–124 (1984).

7. R. Lemmert, *On ordinary differential equations in locally convex space*. Nonlinear Anal. **10**, 1385–1390 (1986).

8. R.H. Martin, *Nonlinear operators and differential equations in Banach spaces*. Wiley, New York (1976).

9. V.M. Millionščikov, *A contribution to the theory of differential equations* $\frac{dx}{dt} = f(x,t)$ *in locally convex spaces*. Dokl. Akad. Nauk SSSR **131**, 510–514 (1960).

10. N.H. Pavel, *Differential equations, flow invariance and applications*. Pitman, Boston - London - Melbourne (1984).

11. A.N. Tychonoff, *Ein Fixpunktsatz*. Math. Ann. **111**, 767–776 (1935).

12. A.N. Tychonoff, *Über unendliche Systeme von Differentialgleichungen*. Mat. Sbornik **41**, 551–555 (1935).

Third International Colloquium on Differential Equations pp. 95-104 (1993)
D. Bainov and V. Covachev (Eds)

ENTIRE SOLUTIONS OF NONLINEAR PARABOLIC EQUATIONS

G. N. Hile, C. P. Mawata

University of Hawaii. Honolulu, HI, USA.
University of Tennessee. Chattanooga, TE, USA.

ABSTRACT

We study entire solutions (i.e., solutions defined in all of $\mathbb{R}^{n+1}$) of second order nonlinear parabolic equations with linear principal part. Under appropriate hypotheses, we establish existence and uniqueness of an entire solution vanishing at infinity. More generally, we discuss existence and uniqueness of an entire solution approaching a given heat polynomial at infinity.

1. INTRODUCTION

We consider solutions in $\mathbb{R}^{n+1} = \mathbb{R}^n \times \mathbb{R}$ of nonlinear parabolic partial differential equations of the general form

$$a(x,t) \cdot \mathcal{D}_x^2 u(x,t) + \mathcal{H}\big(x,t,u(x,t),\mathcal{D}_x u(x,t)\big) - u_t(x,t) = 0 \quad ; \tag{1}$$

such a solution defined in all of $\mathbb{R}^{n+1}$ we call an *entire* solution. The $n \times n$ matrix $a(x,t)$ has real entries, is symmetric and positive definite, and approaches the identity matrix at infinity; it is also Hölder continuous with Hölder constant decaying at an appropriate rate at infinity. The real valued function $\mathcal{H} = \mathcal{H}(x,t,u,z)$, defined on all of $\mathbb{R}^{2n+1} = \mathbb{R}^n \times \mathbb{R} \times \mathbb{R} \times \mathbb{R}^n$, is measurable in the first two variables (x,t) and Lipschitz continuous in the last two variables (u,z), with Lipschitz constant decaying at an appropriate rate at infinity. We consider solutions of (1) in the space $W^{2,1,p}_{loc}(\mathbb{R}^{n+1})$, with $p > n+2$; in this space functions themselves are continuous, and hence pointwise behaviour near infinity may be discussed. We show that, under certain hypotheses, there exists a unique entire solution of (1) vanishing at infinity. More generally, under stronger hypotheses, given a polynomial solution P of the heat equation with certain allowable growth rate, there exists a unique solution u of (1) such that $u - P$ vanishes at infinity.

One consequence of our results is a Liouville theorem for the homogeneous linear parabolic equation,

$$a(x,t) \cdot \mathcal{D}_x^2 u(x,t) + b(x,t) \cdot \mathcal{D}_x u(x,t) - u_t(x,t) = 0 \quad .$$

Under certain continuity assumptions on a and other restrictions on the behaviour of the coefficients a, b, and c at infinity, we prove that any entire and bounded solution of this equation in the space $W^{2,1,p}_{loc}(\mathbb{R}^{n+1})$ is necessarily constant. This classical Liouville theorem holds also for the nonlinear equation (1), again under certain conditions; we require however that constant functions solve the equation.

For related work on Liouville-type theorems for partial differential equations, see [1-18].

Because of space considerations we do not give complete proofs; a more detailed presentation will appear elsewhere.

2. L^p ESTIMATES FOR THE LINEAR EQUATION

We denote a point in $\mathbb{R}^{n+1} = \mathbb{R}^n \times \mathbb{R}$, $n \geq 1$, as $(x,t) = (x_1, x_2, \ldots, x_n, t)$, where $x \in \mathbb{R}^{n+1}$, $t \in \mathbb{R}$. We consider a linear differential operator $\mathbb{P}$ with real valued and measurable coefficients defined in all of $\mathbb{R}^{n+1}$,

$$\begin{aligned} \mathcal{P}u &= a \cdot \mathbb{D}_x^2 u + b \cdot \mathbb{D}_x u + c\, u - u_t \\ &= \sum_{i,j=1}^{n} a_{ij}\, \partial_{ij} u + \sum_{i=1}^{n} b_i\, \partial_i u + c\, u - \partial_{n+1} u \quad . \end{aligned} \tag{2}$$

Here $a = (a_{ij})$ is a real symmetric $n \times n$ matrix, $b = (b_i)$ a real n-vector, and c a real scalar, while the real valued function $u = u(x,t)$ is to be defined on all of $\mathbb{R}^{n+1}$. The n-vector $\mathbb{D}_x u = (\partial_i u)_{n\times 1}$ represents the first order *space derivatives* of u, $\partial_i u := \partial u/\partial x_i$, $1 \leq i \leq n$, and the $n \times n$ matrix $\mathbb{D}_x^2 u = (\partial_{ij} u)_{n\times n}$ consists of all space derivatives of u of second order, $\partial_{ij} u := \partial^2 u/\partial x_i \partial x_j$, $1 \leq i,j \leq n$; these derivatives are understood in the Sobolev sense. All coefficients a_{ij}, b_i, c likewise are functions of (x,t) in $\mathbb{R}^{n+1}$.

For economy of notation we introduce the quantity

$$\rho(x,t) := \sqrt{1 + |x|^2 + |t|} \quad ,$$

and for $\sigma \in \mathbb{R}$, $1 \leq p \leq \infty$, we utilize the weighted L^p norms

$$\|\rho^\sigma F\|_p := \begin{cases} \left[\int_{\mathbb{R}^{n+1}} \rho(x,t)^{\sigma p}\, |F(x,t)|^p \, dx\, dt \right]^{1/p} , & 1 \leq p < \infty \\ \operatorname*{ess\,sup}_{(x,t) \in \mathbb{R}^{n+1}} \rho(x,t)^\sigma\, |F(x,t)| , & p = \infty . \end{cases}$$

Note that these norms increase with σ when p is held fixed.

We assume the following conditions on the coefficients of $\mathcal{P}$ (I is the $n\times n$ identity matrix) :

(P1) There exists a positive constant λ such that for all (x,t) in $\mathbb{R}^{n+1}$ and ξ in $\mathbb{R}^n$,

$$a(x,t)\xi \cdot \xi \geq \lambda\ |\xi|^2 .$$

(P2) There exist constants α, K, with $0 < \alpha \leq 1$, $K \geq 0$, such that for all (x,t) and (y,s) in $\mathbb{R}^{n+1}$,

$$|a(x,t) - a(y,s)| \leq K \left[\frac{|x-y|^2 + |t-s|}{1 + |x|^2 + |y|^2 + |t| + |s|} \right]^{\alpha/2} .$$

(P3) Both b and c are measurable in $\mathbb{R}^{n+1}$, and there exist nonnegative constants Λ and γ such that

$$\|\rho^{\gamma}(a-I)\|_\infty ,\ \|\rho^{1+\gamma}b\|_\infty ,\ \|\rho^{2+\gamma}c\|_\infty \leq \Lambda .$$

Definition For Q an open subset of $\mathbb{R}^{n+1}$, the space $W^{2,1,p}(Q)$ consists of those functions $u = u(x,t)$, defined in Q, such that u, $\mathbb{D}_x u$, $\mathbb{D}_x^2 u$, u_t are in $L^p(Q)$. We say $u \in W^{2,1,p}_{loc}(Q)$ if $u \in W^{2,1,p}(Q')$ for every open subset Q' of Q whose closure lies in Q.

Using analogous results for bounded domains, as found for example in [19], chapter IV, we first establish the following two lemmas concerning the operator $\mathcal{P}$ and functions u in $W^{2,1,p}_{loc}(\mathbb{R}^{n+1})$:

Lemma 1A Assume conditions (P1), (P2), (P3) on the coefficients of $\mathcal{P}$, let $u \in W^{2,1,p}_{loc}(\mathbb{R}^{n+1})$ with $1 < p < \infty$, and let $\sigma \in \mathbb{R}$; then

$$\|\rho^{\sigma+1}\ \mathbb{D}_x u\|_p + \|\rho^{\sigma+2}\ \mathbb{D}_x^2 u\|_p + \|\rho^{\sigma+2}\ u_t\|_p$$

$$\leq C(n,p,\Lambda,\lambda,K,\alpha,\sigma)\ \left[\ \|\rho^{\sigma}\ u\|_p + \|\rho^{\sigma+2}\ \mathcal{P}u\|_p\ \right] . \qquad \blacksquare$$

Lemma 1B If $n+2 < p < \infty$ and $\sigma \in \mathbb{R}$, then for $u \in W^{2,1,p}_{loc}(\mathbb{R}^{n+1})$,

$$\|\rho^{\sigma+(n+2)/p}\ u\|_\infty + \|\rho^{\sigma+1+(n+2)/p}\ \mathbb{D}_x u\|_\infty$$

$$\leq C(n,p,\sigma)\ \left[\ \|\rho^{\sigma}u\|_p + \|\rho^{\sigma+2}\ \mathbb{D}_x^2 u\|_p + \|\rho^{\sigma+2}\ u_t\|_p\ \right] . \qquad \blacksquare$$

We let Δu denote the Laplacian of u with respect to the space variables; thus $\Delta u = \partial_{11}u + \ldots + \partial_{nn}u$. Combining our lemmas, we arrive at the following fundamental L^p estimate concerning the operator $\mathcal{P}$.

Theorem 1 Assume conditions (P1), (P2), (P3) on the coefficients of $\mathcal{P}$, let $u \in W^{2,1,p}_{loc}(\mathbb{R}^{n+1})$ with $n+2 < p < \infty$, and let $\sigma \in \mathbb{R}$; then

$$\|\rho^{\sigma+1}\,\mathbb{D}_x u\|_p + \|\rho^{\sigma+2}\,\mathbb{D}_x^2 u\|_p + \|\rho^{\sigma+2}\,u_t\|_p$$

$$+ \|\rho^{\sigma+(n+2)/p}\,u\|_\infty + \|\rho^{\sigma+1+(n+2)/p}\,\mathbb{D}_x u\|_\infty$$

$$\le C(n,p,\Lambda,\lambda,K,\alpha,\sigma)\,\left[\,\|\rho^{\sigma}\,u\|_p + \|\rho^{\sigma+2}\,\mathcal{P}u\|_p\right] ,$$

and

$$\|\rho^{\sigma+2}\,(\Delta u-u_t)\|_p \le C(n,p,\Lambda,\lambda,K,\alpha,\sigma-\gamma)\,\left[\,\|\rho^{\sigma-\gamma}\,u\|_p + \|\rho^{\sigma+2}\,\mathcal{P}u\|_p\,\right] .$$

∎

3. BOUNDS FOR THE NONHOMOGENEOUS HEAT EQUATION

We discuss now the nonhomogeneous heat equation in $\mathbb{R}^{n+1} = \mathbb{R}^n \times \mathbb{R}$,

$$\Delta u - u_t = f \quad . \tag{3}$$

We require the so-called *fundamental solution* of the heat equation in $\mathbb{R}^{n+1}$,

$$\mathbb{K}(x,t) := \begin{cases} (4\pi t)^{-n/2}\, e^{-|x|^2/4t} & , \text{ if } t > 0 \\ 0 & , \text{ if } t \le 0 \end{cases} .$$

For appropriate functions f, we investigate the integral $\mathcal{K}f$ defined as

$$\mathcal{K}f(x,t) := \int_{\mathbb{R}^{n+1}} \mathbb{K}(x-y,t-s)\, f(y,s)\, dy\, ds \quad .$$

Using properties of the fundamental solution $\mathbb{K}$, along with some intricate bounds concerning the operator $\mathcal{K}$, we arrive at the following apriori bound on entire solutions, vanishing at infinity, of the nonhomogeneous heat equation :

Theorem 2 Suppose

$$n+2 < p < \infty \quad , \quad -\frac{n+2}{p} < \sigma < \infty ,$$

and let f be a real valued measurable function on $\mathbb{R}^{n+1}$ with $\|\rho^{\sigma+2}\, f\|_p$ finite. Then there exists exactly one solution in $W^{2,1,p}_{loc}(\mathbb{R}^{n+1})$ and vanishing at infinity of (3); this solution is $u = -\mathcal{K}f$, and whenever

$$-\frac{n+2}{p} < \sigma < n - \frac{n+2}{p}$$

we have the bound

$$\|\rho^{\sigma}\,\mathcal{K}f\|_p + \|\rho^{\sigma+1}\,\mathbb{D}_x(\mathcal{K}f)\|_p + \|\rho^{\sigma+2}\,\mathbb{D}_x^2(\mathcal{K}f)\|_p + \|\rho^{\sigma+2}\,(\mathcal{K}f)_t\|_p$$

$$+ \|\rho^{\sigma+(n+2)/p}\,\mathcal{K}f\|_\infty + \|\rho^{\sigma+1+(n+2)/p}\,\mathbb{D}_x(\mathcal{K}f)\|_\infty$$

$$\le C(n,p,\sigma)\,\|\rho^{\sigma+2}\,f\|_p \quad .$$

∎

4. ENTIRE SOLUTIONS OF THE LINEAR EQUATION

We return to the linear operator $\mathcal{P}$ of §2, and establish an apriori bound stronger than that of Theorem 1, but for functions u vanishing at infinity. We assume again conditions (P1)-(P3) on the coefficients of $\mathcal{P}$, but in (P3) we require that $\gamma > 0$; thus $\mathcal{P}$ approaches the heat operator, $\Delta - \partial_t$, at infinity. We discuss solutions u in $W^{2,1,p}_{loc}(\mathbb{R}^{n+1})$ with $n+2 < p < \infty$; such functions are continuous in $\mathbb{R}^{n+1}$, by Lemma 1C.

Theorem 3 Assume conditions (P1)-(P3) on the coefficients of $\mathcal{P}$, with also $\gamma > 0$ and $c \le 0$ in $\mathbb{R}^{n+1}$. Assume moreover that

$$n+2 < p < \infty \quad , \quad -\frac{n+2}{p} < \sigma < n - \frac{n+2}{p} \quad .$$

If $u \in W^{2,1,p}_{loc}(\mathbb{R}^{n+1})$ and $u(\infty) = 0$, then

$$\begin{aligned} &\|\rho^{\sigma} u\|_p + \|\rho^{\sigma+1} \mathfrak{D}_x u\|_p + \|\rho^{\sigma+2} \mathfrak{D}^2_x u\|_p + \|\rho^{\sigma+2} u_t\|_p \\ &\le C(n,p,\Lambda,\lambda,K,\alpha,\sigma,\gamma) \|\rho^{\sigma+2} \mathcal{P}u\|_p \quad . \end{aligned} \tag{4}$$

Outline of Proof We may assume that $\|\rho^{\sigma+2} \mathcal{P}u\|_p$ is finite, as otherwise (4) is clear. Using bounds of §2, we show first that $\|\rho^{\sigma} u\|_p$ is finite. In order to establish (4), by Theorem 1 we need only justify the bound

$$\|\rho^{\sigma} u\|_p \le C(n,p,\Lambda,\lambda,K,\alpha,\sigma,\gamma) \|\rho^{\sigma+2} \mathcal{P}u\|_p \quad , \tag{5}$$

for all u in $W^{2,1,p}_{loc}(\mathbb{R}^{n+1})$ with $u(\infty) = 0$ and $\|\rho^{\sigma+2} \mathcal{P}u\|_p$ finite. If (5) does not hold for such functions, then there exist sequences $\{a_m\}$, $\{b_m\}$, $\{c_m\}$, $\{u_m\}$, $m = 1, 2, 3, \ldots$, of real valued functions in $\mathbb{R}^{n+1}$ such that $\{a_m\}$, $\{b_m\}$, $\{c_m\}$ satisfy conditions (P1)-(P3) with $c_m \le 0$, each $u_m \in W^{2,1,p}_{loc}(\mathbb{R}^{n+1})$ with $u_m(\infty) = 0$, and

$$\|\rho^{\sigma} u_m\|_p = 1 \quad , \quad \|\rho^{\sigma+2} \mathcal{P}_m u_m\|_p \le \frac{1}{m} \quad , \tag{6}$$

where $\mathcal{P}_m$ is defined as the operator

$$\mathcal{P}_m u = a_m \cdot \mathfrak{D}^2_x u + b_m \cdot \mathfrak{D}_x u + c_m u - u_t \quad .$$

By compactness arguments we can demonstrate that a subsequence may be chosen, which we relabel as the original sequence, so that for any bounded open set Q in $\mathbb{R}^{n+1}$ we have in $L^p(Q)$ the weak limits

$$a_m \cdot \mathfrak{D}^2_x u_m \longrightarrow a \cdot \mathfrak{D}^2_x u \;, \quad b_m \cdot \mathfrak{D}_x u_m \longrightarrow b \cdot \mathfrak{D}_x u \;, \quad c_m \cdot u_m \longrightarrow c\, u \;,$$
$$\partial u_m/\partial t \longrightarrow \partial u/\partial t \;, \quad \mathcal{P}_m u_m \longrightarrow 0 \;.$$

Since weak limits are uniquely determined almost everywhere, u is a solution in $W^{2,1,p}_{loc}(\mathbb{R}^{n+1})$ of the homogeneous equation

$$a \cdot \mathfrak{D}_x^2 u + b \cdot \mathfrak{D}_x u + c\, u - u_t = 0 \quad .$$

We conclude also that $u(\infty) = 0$ and then, by a maximum principle of Tso [20], that $u \equiv 0$. This observation leads to a contradiction of (6). ∎

5. THE NONLINEAR EQUATION

We consider the nonlinear partial differential equation

$$a(x,t) \cdot \mathfrak{D}_x^2 u(x,t) + \mathcal{H}\big(x,t,u(x,t),\mathfrak{D}_x u(x,t)\big) - u_t(x,t) = 0 \quad . \tag{7}$$

The function $\mathcal{H}$ is real valued on $\mathbb{R}^n \times \mathbb{R} \times \mathbb{R} \times \mathbb{R}^n = \mathbb{R}^{2n+2}$; we shall write $\mathcal{H} = \mathcal{H}(x,t,u,z)$, where $x \in \mathbb{R}^n$, $t \in \mathbb{R}$, $u \in \mathbb{R}$, $z \in \mathbb{R}^n$. We assume the following conditions on the $n \times n$ real symmetric matrix a and the function $\mathcal{H}$:

(N1) There exists $\lambda > 0$ such that for all (x,t) in $\mathbb{R}^{n+1}$ and ξ in $\mathbb{R}^n$,

$$a(x,t)\xi \cdot \xi \geq \lambda\, |\xi|^2 \quad .$$

(N2) There exist α, K with $0 < \alpha \leq 1$, $K \geq 0$, such that for all (x,t) and (y,s) in $\mathbb{R}^{n+1}$,

$$|a(x,t) - a(y,s)| \leq K \left[\frac{|x-y|^2 + |t-s|}{1 + |x|^2 + |y|^2 + |t| + |s|} \right]^{\alpha/2} \quad .$$

(N3) There exist Λ, γ, with $\Lambda \geq 0$, $\gamma > 0$, such that for $(x,t) \in \mathbb{R}^{n+1}$,

$$|a(x,t) - I| \leq \Lambda\, \rho(x,t)^{-\gamma} \quad ,$$

and for all $(x,t) \in \mathbb{R}^{n+1}$, $u,v \in \mathbb{R}$, $z,\zeta \in \mathbb{R}^n$,

$$|\mathcal{H}(x,t,u,z) - \mathcal{H}(x,t,u,\zeta)| \leq \Lambda\, |z-\zeta|\, \rho(x,t)^{-1-\gamma} \quad ,$$

$$|\mathcal{H}(x,t,u,z) - \mathcal{H}(x,t,v,z)| \leq \Lambda\, |u-v|\, \rho(x,t)^{-2-\gamma} \quad .$$

(N4) For all $(x,t) \in \mathbb{R}^{n+1}$, $z \in \mathbb{R}^n$, and $u,v \in \mathbb{R}$ with $u \neq v$,

$$\frac{\mathcal{H}(x,t,u,z) - \mathcal{H}(x,t,v,z)}{u - v} \leq 0 \quad .$$

(N5) For each $(u,z) \in \mathbb{R} \times \mathbb{R}^n$, $\mathcal{H}(\cdot,*,u,z)$ is measurable as a function of (x,t) on $\mathbb{R}^{n+1}$; moreover, there exist numbers p and σ such that

$$n+2 < p < \infty \quad , \quad -\frac{n+2}{p} < \sigma < \infty \ ,$$

$$\|\rho^{\sigma+2}\, \mathcal{H}(\cdot,*,0,0)\|_p < \infty \ .$$

In the course of the "linearization" of equation (7), we employ the notation

$$C(x,t,u,v,z) = \begin{cases} \dfrac{\mathcal{H}(x,t,u,z) - \mathcal{H}(x,t,v,z)}{u - v} & , \text{ if } u \neq v \\ 0 & , \text{ if } u = v \ , \end{cases}$$

$$z_{\cdot i} = \begin{cases} (z_1, z_2, \dots, z_i, 0, \dots, 0) & , \text{ if } i = 1, 2, \dots, n \\ (0, \dots, 0) & , \text{ if } i = 0 \end{cases} ,$$

and for $i = 1, 2, \dots, n$,

$$\mathcal{B}_i(x,t,u,z,\zeta) = \begin{cases} \dfrac{\mathcal{H}\big(x,t,u,\zeta+(z-\zeta)_{\cdot i}\big) - \mathcal{H}\big(x,t,u,\zeta+(z-\zeta)_{\cdot i-1}\big)}{z_i - \zeta_i} & , z_i \neq \zeta_i \\ 0 & , z_i = \zeta_i . \end{cases}$$

Note that conditions (N3) and (N4) yield the inequalities

$$|\mathcal{B}_i(x,t,u,z,\zeta)| \le \Lambda \, \rho(x,t)^{-1-\gamma} ,$$

$$|C(x,t,u,v,z)| \le \Lambda \, \rho(x,t)^{-2-\gamma} ,$$

$$C(x,t,u,v,z) \le 0 ,$$

and that, if we introduce a vector

$$\mathcal{B} = (\mathcal{B}_1, \mathcal{B}_2, \dots, \mathcal{B}_n) ,$$

we may write

$$\begin{aligned} \mathcal{H}(x,t,u,z) - \mathcal{H}(x,t,v,\zeta) &= \sum_{i=1}^{n} \mathcal{B}_i(x,t,v,z,\zeta)(z_i-\zeta_i) + C(x,t,u,v,z)(u-v) \\ &= \mathcal{B}(x,t,v,z,\zeta) \cdot (z-\zeta) + C(x,t,u,v,z)\,(u-v) . \end{aligned}$$

The nonlinear equation (7) may be written as a "linearized equation",

$$\mathcal{Q}w := a(\cdot,*) \cdot \mathcal{D}_x^2 w + \mathcal{B}(\cdot,*,v,\mathcal{D}_x u,\mathcal{D}_x v) \cdot \mathcal{D}_x w + C(\cdot,*,u,v,\mathcal{D}_x u)\, w - w_t = 0 .$$

Using the bounds of §4 for the linear equation, we use a Schauder continuation argument to establish the following existence and uniqueness result for the nonlinear equation :

Theorem 4 Assume conditions (N1)-(N5) on $\mathcal{H}$ and the matrix a. Then there exists in the space $W^{2,1,p}_{loc}(\mathbb{R}^{n+1})$ a unique entire solution u of (7) such that $u(\infty) = 0$; if moreover $\sigma < n - (n+2)/p$, then for this solution we have the bound

$$\begin{aligned} &\|\rho^{\sigma} u\|_p + \|\rho^{\sigma+1} \mathcal{D}_x u\|_p + \|\rho^{\sigma+2} \mathcal{D}_x^2 u\|_p + \|\rho^{\sigma+2} u_t\|_p \\ &\le C(n,p,\Lambda,\lambda,K,\alpha,\sigma,\gamma) \, \|\rho^{\sigma+2} \mathcal{H}(\cdot,*,0,0)\|_p . \end{aligned}$$

From Theorem 4 we obtain the following more general result, where u is allowed polynomial growth at infinity. Recall that a *heat polynomial* is any polynomial solution of the heat equation, $\Delta u - u_t = 0$.

Theorem 5 Assume conditions (N1)-(N5) on $\mathcal{H}$ and the matrix a.

(a) For any heat polynomial P of degree less than γ in x, there exists in the space $W^{2,1,p}_{loc}(\mathbb{R}^{n+1})$ a unique entire solution of (7) such that the difference u - P vanishes at infinity.

(b) If u is an entire solution of (7) in the space $W^{2,1,p}_{loc}(\mathbb{R}^{n+1})$, and

$$|u(x,t)| = O\big(\rho(x,t)^{\tau}\big) \quad \text{as } (x,t) \longrightarrow \infty$$

for some τ, $0 \le \tau < \gamma$, then there exists a unique heat polynomial $P = P(x,t)$, of degree no larger that τ in x and no larger than $\tau/2$ in t, such that u - P vanishes at infinity. ∎

A classical "Liouville Theorem" states that a bounded entire solution of an equation must be constant. For such a theorem to be possible, constant functions must solve the equation. For equation (7), we see that this requirement is equivalent to

$$\mathcal{H}(x,t,c,0) = 0 \quad , \text{ for all } (x,t) \in \mathbb{R}^{n+1} \text{ and } c \in \mathbb{R} . \tag{8}$$

Corollary 5A Assume conditions (N1)-(N5) on a and $\mathcal{H}$, and suppose (8) holds. Then any bounded and entire solution of (7) in $W^{2,1,p}_{loc}(\mathbb{R}^{n+1})$ is constant.

Proof If u is such an entire and bounded solution, then Theorem 5(b), applied with $\tau = 0$, asserts that for some constant c the difference u - c vanishes at infinity. But the constant solution c has this property; by the uniqueness statement of Theorem 5(a), $u \equiv c$. ∎

Finally, we specialize to the linear equation,

$$\mathcal{L}u = a \cdot \mathfrak{D}^2_x u + b \cdot \mathfrak{D}_x u + c\,u - u_t + f = 0 \quad , \tag{9}$$

which has the form of (7) if we take $\mathcal{H}$ as

$$\mathcal{H}(x,t,u,z) = b(x,t) \cdot z + c(x,t)\,u + f(x,t) \quad .$$

If we assume conditions (P1)-(P3) of §2 on a, b, and c, with $\gamma > 0$ and $c \le 0$, and with $\|\rho^{\sigma+2} f\|_p$ finite where $\sigma > -(n+2)/p$, then conditions (N1)-(N5) are easily verified for a and $\mathcal{H}$. Therefore, under these conditions, Theorems 4 and 5 apply to equation (9). In particular, when $f \equiv 0$ we have the linear homogeneous equation

$$\mathcal{P}u = a \cdot \mathfrak{D}^2_x u + b \cdot \mathfrak{D}_x u + c\,u - u_t = 0 \quad ; \tag{10}$$

and when also $c \equiv 0$, constant functions solve this equation, allowing application of the Liouville Theorem of Corollary 5A.

Entire solutions of (10) form a real vector space; and theorem 5 yields immediately the following :

Theorem 6 Suppose conditions (P1)-(P3) apply to the coefficients of $\mathcal{P}$ as defined by (10), with also $c \leq 0$ in $\mathbb{R}^{n+1}$ and $\gamma > 0$. If $0 \leq \tau < \gamma$, then the space of entire solutions of $\mathcal{P}u = 0$ in $W^{2,1,p}_{loc}(\mathbb{R}^{n+1})$, with the property that

$$|u(x,t)| = O(\rho(x,t)^{\tau}) \text{ as } (x,t) \longrightarrow \infty ,$$

form a finite dimensional real vector space, of dimension the same as the space of heat polynomials of degree in x no larger than τ. There is a one-to-one correspondence between the members of these spaces, as described in the statement of Theorem 5. ▮

REFERENCES

1. H. Begehr and G. N. Hile, Schauder estimates and existence theory for entire solutions of linear elliptic equations, Proc. Roy. Soc. Edinburgh 110A (1988), 101-123.
2. E. Bohn and L. K. Jackson, The Liouville theorem for a quasilinear elliptic partial differential equation, Trans. Amer. Math. Soc. 104 (1962), 392-397.
3. A. Friedman, Bounded entire solutions of elliptic equations, Pacific J. Math. 44 (1973), 497-507.
4. D. Gilbarg and J. Serrin, On isolated singularities of solutions of second order elliptic differential equations, J. Analyse Math. 4 (1954-6), 309-340.
5. S. Hildebrandt and K. Widman, Sätze vom Liouvilleschen Typ für quasilineare elliptische Gleichungen und Systeme, Nachr. Akad. Wiss. Göttingen Math. Phys. Kl. II (1979), Nr. 4, 41-59.
6. G. N. Hile, Entire solutions of linear elliptic equations with Laplacian principal part, Pacific J. Math. 62 (1976), 127-140.
7. G. N. Hile and C. P. Mawata, Existence and uniqueness of solutions to quasilinear elliptic equations, Proceedings of the International Conference on the Theory and Applications of Differential Equations, 1991.
8. G. N. Hile and C. P. Mawata, Liouville theorems for nonlinear elliptic equations of second order, to appear.
9. A. V. Ivanov, Local estimates for the first derivatives of solutions of quasilinear second order elliptic equations and their application to Liouville type theorems, Sem. Steklov Math. Inst. Leningrad 30 (1972), 40-50; Translated in J. Sov. Math. 4 (1975), 335-344.
10. C. P. Mawata, Schauder estimates and existence theory for entire solutions of linear parabolic equations, Differential and Integral Equations 2, No. 3 (1989), 251-274.
11. L. Nirenberg and H. F. Walker, The null spaces of elliptic partial differential operators in $\mathbb{R}^n$, J. Math. Anal. Appl. 42 (1973), 271-301.
12. L. Peletier and J. Serrin, Gradient bounds and Liouville theorems for quasilinear elliptic equations, Ann. Scuola Norm. Sup. Pisa Cl. Sci., Ser. 4, 5 (1978), 65-104.
13. R. Redheffer, On the inequality $\Delta u \geq f(u,|\text{grad } u|)$, J. Math. Anal. 1 (1960), 277-299.

14. J. Serrin, Entire solutions of nonlinear Poisson equations, Proc. London Math. Soc. 24 (1972), 348-366.
15. J. Serrin, Liouville theorems and gradient bounds for quasilinear elltiptic systems, Arch. Rational Mech. Anal. 66 (1977), 295-310.
16. H. F. Walker, On the null-spaces of first-order elliptic partial differential operators in $\mathbb{R}^n$, Proc. Amer. Math. Soc. 30 (1971), 278-286.
17. H. F. Walker, On the null-spaces of elliptic partial differential operators in $\mathbb{R}^n$, Trans. Amer. Math. Soc. 173 (1972), 263-275.
18. N. Weck, Liouville theorems for linear elliptic systems, Proc. Roy. Soc. Edinburgh 94A (1983), 309-322.
19. O. A. Ladyzenskaja, V. A. Solonnikov, and N. N. Ural'ceva, Linear and Quasilinear Equations of Parabolic Type, Amer. Math. Soc. Translations of Mathematical Monographs 23, 1968.
20. Kaising Tso, On an Aleksandrov-Bakel'man type maximum principle for second order parabolic equations, Comm. Partial Diff. Eqns. 10 (1985), 543-553.

Third International Colloquium on Differential Equations pp. 105-117 (1993)
D. Bainov and V. Covachev (Eds)

A Class of Weighted-Mean Banach Spaces

Tepper L. Gill and W.W. Zachary
Computational Science and Engineering Research Center
Howard University
Washington, D.C. 20059 USA

ABSTRACT

We introduce a new class of Banach spaces $F_p(\Omega)$, $1 \leq p \leq \infty$, with Ω an open subset of $\mathbb{R}^n$ $(n \geq 1)$, which contain the respective Lebesgue spaces $L^p(\Omega)$ as dense continuous embeddings. Results are stated concerning various properties of these spaces including separability properties, duality properties, and compactness of L^p embeddings into F_p when $1 < p < \infty$. Of particular interest is the fact that these spaces contain the Denjoy integrable functions and the Schwartz distributions under appropriate conditions. The F_p spaces depend upon a denumerable set of auxiliary functions. Several possible choices for these functions are discussed and properties of the corresponding F_p spaces are investigated.

1. INTRODUCTION

Let $H = L^2(\mathbb{R}^n)$ and suppose that $\{e_m, m = 1,2,\ldots\}$ is a complete orthonormal set in H. By the Parseval relation, for each $f \in H$ we have

$$\left[\int_{\mathbb{R}^n} |f|^2 dx\right]^{1/2} = \left[\sum_{m=1}^{\infty} \left|\int_{\mathbb{R}^n} f\, \overline{e_m}\, dx\right|^2\right]^{1/2}. \tag{1.1}$$

Suppose that $f \notin H$, e.g. $f(x) = \exp(ix^2)$. Then the left-hand side of (1.1) is infinite, but each term on the right-hand side is finite if each $e_m \in L^2(\mathbb{R}^n) \cap L^1(\mathbb{R}^n)$. This situation suggests that it may be possible to define larger function spaces than L^2 by using expressions analogous to the right-hand side of (1.1) to define appropriate norms.

We will discuss how an elaboration of this idea leads to the definition of a new class of Banach spaces, called F_p spaces, which contain the respective L^p spaces. They have many properties which make them ideal for the study of nonlinear problems involving highly oscillatory functions.

Let Ω be an open subset of $\mathbb{R}^n$ and suppose that $\{t_\gamma, \gamma \in \Gamma\}$ is a set of positive real numbers such that

$$(1.2) \qquad \sum_{\gamma\in\Gamma} t_\gamma = 1.$$

For $p \in [1,\infty]$ and $p^{-1} + q^{-1} = 1$, let $\{e_\gamma, \gamma \in \Gamma\} \subset L^q(\Omega)$ be a *total family for* $L^p(\Omega)$, i.e.,

$$(1.3) \qquad \begin{cases} f \in L^p(\Omega) \text{ and } (e_\gamma, f) = \int_\Omega f(x)\overline{e_\gamma(x)}dx = 0 \\ \text{for all } \gamma \in \Gamma \text{ implies that } f = 0 \text{ a.e. on } \Omega. \end{cases}$$

We define the quantities:

$$(1.4) \qquad \|f\|_{F_p(\Omega)} = \begin{cases} \left(\sum_{\gamma\in\Gamma} t_\gamma \, |(e_\gamma, f)|^p\right)^{1/p}, & 1 \le p < \infty, \\ \sup_{\gamma\in\Gamma} |(e_\gamma, f)|, & p = \infty. \end{cases}$$

The following result is an easy consequence of the Hölder inequalities for L^p spaces.

<u>LEMMA 1.1</u>. <u>Assuming (1.2), (1.3) and using (1.4), define the numbers</u>

$$a_p = \begin{cases} \left(\sum_{\gamma\in\Gamma} t_\gamma \|e_\gamma\|^p_{L^q(\Omega)}\right)^{1/p}, & 1 \le p < \infty, \\ \sup_{\gamma\in\Gamma} \|e_\gamma\|_{L^1(\Omega)}, & p = \infty. \end{cases}$$

<u>If</u> $a_p < \infty$, <u>then</u> $\|f\|_{F_p(\Omega)} \le a_p \|f\|_{L^p(\Omega)}$, $f \in L^p(\Omega)$.

Thus, under the hypotheses of Lemma 1.1, the quantities (1.4) are finite for $f \in L^p(\Omega)$. Furthermore, it is easily shown that these quantities are norms on $L^p(\Omega)$, $1 \le p \le \infty$. However, because of condition (1.2), $L^p(\Omega)$ are not complete relative to the norms (1.4). We define spaces $F_p(\Omega)$ as the respective completions of $L^p(\Omega)$ with respect to the norms $\|\cdot\|_{F_p(\Omega)}$ for each $p \in [1,\infty]$ for which the conditions in Lemma 1.1 are satisfied. Each F_p is a Banach space which contains the corresponding L^p as a dense subspace and the embedding $L^p \hookrightarrow F_p$ is continuous. F_p-type spaces in which the norms contain derivatives (analogous to Sobolev spaces) involve a higher level of complication. We will discuss this situation in Section 3.

The F_p spaces are easy to construct, are separable, and contain, for appropriate choices of the auxiliary functions $\{e_\gamma, \gamma \in \Gamma\}$, the Schwartz distributions and the Denjoy integrable functions [1,3]. In this connection, we note that there are several types of Denjoy integrals, each of which is a generalization of both the Riemann and Lebesgue integrals.

Our results in Theorem 2.2(iii), (v) are important for applications to nonlinear partial differential equations because they imply that weakly convergent sequences in L^p spaces are strongly convergent in the corresponding F_p spaces. Therefore, the use of F_p spaces should have some advantages over current approaches that use weak convergence in L^p spaces [4] because the latter spaces cannot handle highly oscillatory integrals directly, but must rely on the existence of cancellations in order to obtain convergence.

Our interest in the F_p spaces arose in work on the foundations of quantum theory associated with the Feynman functional calculus [5,6], and this is the reason for the appellation F_p. We note that a version of F_2 was used by Kuelbs [7] to study Gaussian measures on Banach spaces (see also [8]).

2. SOME PROPERTIES OF THE F_p SPACES.

In this section we will discuss (without proofs) some functional analytic properties of the F_p spaces. It is well known that $L^p(\Omega)$ is separable when $1 \leq p < \infty$ but is not separable when $p = \infty$. However, for $F_p(\Omega)$ we have the following remarkable results:

THEOREM 2.1. *Assume the hypotheses of Lemma 1.1. Then*

(a) $F_p(\Omega)$ *is separable when* $1 \leq p < \infty$.

(b) *If, in addition,* $\{e_\gamma, \gamma \in \Gamma\} \subset L^\infty(\Omega)$ *and*

$$\sum_{\gamma\in\Gamma} t_\gamma \|e_\gamma\|^p_{L^\infty(\Omega)} < \infty, \tag{2.1}$$

then $L^1(\Omega) \subset F_p(\Omega)$ *densely when* $1 \le p < \infty$.

(c) *If the hypotheses of part (b) hold with (2.1) replaced by the condition* $\sup_{\gamma\in\Gamma} \|e_\gamma\|_{L^\infty(\Omega)} < \infty$, *then* $L^1(\Omega) \subset F_\infty(\Omega)$ *densely and* $F_\infty(\Omega)$ *is separable.*

<u>**THEOREM 2.2**</u>. *Assume the hypotheses of Lemma 1.1. Then:*

(i) (*Hölder inequalities*) *For* $p^{-1} + q^{-1} = 1$ *with* $1 \le p, q \le \infty$, *we have* $|\langle f,g\rangle| \le \|f\|_{F_p} \|g\|_{F_q}$ *for each pair* $\{f,g\} \in F_p(\Omega) \times F_q(\Omega)$ *where, for fixed* $f \in F_p$, $\langle f,\cdot\rangle$ *denotes the continuous linear functional* $\sum_{\gamma\in\Gamma} t_\gamma (f,e_\gamma)(e_\gamma,\cdot)$ *on* $F_q(\Omega)$.

(ii) (*Minkowski inequalities*) *If* $1 \le p \le \infty$ *and* $f,g \in F_p(\Omega)$, *then* $\|f + g\|_{F_p} \le \|f\|_{F_p} + \|g\|_{F_p}$.

(iii) *If* K *is a weakly sequentially compact set in* $L^p(\Omega)$, $1 \le p < \infty$, *then* K *is compact in* $F_p(\Omega)$.

(iv) If $1 < p < \infty$, *then* $F_p(\Omega)$ *is uniformly convex.*

(v) *If* $1 < p < \infty$, *then the embeddings* $L^p(\Omega) \hookrightarrow F_p(\Omega)$ *are compact.*

(vi) *If* $1 < p < \infty$ *and* $p^{-1} + q^{-1} = 1$, *then the dual space of* $F_p(\Omega)$ *is isometrically isomorphic to a subspace of* $F_q(\Omega)$.

(vii) *If* $1 \le p < \infty$ *and* $\{e_\gamma, \gamma \in \Gamma\}$ *is bounded in* $L^\infty(\Omega)$ *uniformly in* γ, *then the second dual space of* $L^1(\Omega)$ *is isomorphic to a proper subspace of* $F_p(\Omega)$.

With respect to (vi), we conjecture that the subspace of $F_q(\Omega)$ is not proper.

3. SOME CHOICES FOR $\{e_\gamma, \gamma \in \Gamma\}$

We discuss some choices for the set $\{e_\gamma, \gamma \in \Gamma\}$ which are useful for applications. Let $\{x_j, j = 1,2,\ldots\}$ denote the set of points in $\mathbb{R}^n$ with rational coordinates. We take $\gamma = (j,k) \in \mathbb{N} \times \mathbb{N} = \Gamma$ and let B_{jk} denote the open ball centered at x_j with radius 2^{-k} and I_{jk} the closed hypercube which circumscribes this ball, respectively. Then four convenient choices for $\{e_\gamma, \gamma \in \Gamma\}$ are obtained by setting $e^{(1)}_\gamma(x) = \chi_{B_{jk}}(x)$, $e^{(2)}_{\gamma,\epsilon}(x) = (J_\epsilon * \chi_{B_{jk}})(x)$, $\epsilon > 0$,

(3.1) $$e^{(3)}_\gamma(x) = \chi_{I_{jk}}(x), \text{ and}$$

(3.2) $$e^{(4)}_\gamma(x) = (\tilde{J}_\epsilon * \chi_{I_{jk}})(x), \ \epsilon > 0,$$

where $\chi_{B_{jk}}$, $\chi_{I_{jk}}$ are the characteristic functions of B_{jk} and I_{jk}, respectively, and J_ϵ denotes the mollifier $J_\epsilon(x) = \epsilon^{-n} J(\frac{x}{\epsilon})$, $\epsilon > 0$, where J is a nonnegative element of $C^\infty_0(\mathbb{R}^n)$ with supp J contained in the closed unit ball centered at the origin and $\|J\|_{L^1(\mathbb{R}^n)} = 1$. Similarly, $\tilde{J}_\epsilon$ denotes the mollifier $\tilde{J}_\epsilon(x) = \epsilon^{-n} \prod_{i=1}^{n} j(\frac{x_i}{\epsilon})$, $\epsilon > 0$, with j a nonnegative element of $C^\infty_0(\mathbb{R})$ with supp j contained in the closed interval $[-1,1]$, and $\|j\|_{L^1(\mathbb{R})} = 1$. The corresponding t_γ are chosen to be independent of k and the summation over $\gamma = (j,k) \in \Gamma = \mathbb{N} \times \mathbb{N}$ is replaced by summation over $j \in \mathbb{N}$. It follows from known results concerning mollifiers that $e^{(i)}_{\gamma,\epsilon} \in C^\infty_0(\Omega)$ for $i = 2$ ($i = 4$) if $\bar{B}_{jk} \subset \Omega$ ($I_{jk} \subset \Omega$) and $\epsilon < \operatorname{dist}(\bar{B}_{jk},\partial\Omega)$ ($\epsilon < \operatorname{dist}(I_{jk},\partial\Omega)$).

The choices (3.1), (3.2) are more convenient for discussion of the relationship between F_p spaces and Denjoy-type integrals in dimensions $n \geq 2$ due to the fact that many higher dimensional nonabsolute integrals are only defined on products of intervals. Other possibilities for $\{e_\gamma, \gamma \in \Gamma\}$ are L^2-orthogonal bases of eigenfunctions of unbounded linear partial differential operators with purely discrete spectra. A different type of choice consists of taking orthonormal wavelet bases for $L^2(\mathbb{R}^n)$ [9].

PROPOSITION 3.1. *Let* Ω *denote an open subset in* $\mathbb{R}^n$. *Then* (a) *the sets* $\{e_{jk}^{(i)};\ i = 1,3,\ j,k \in \mathbb{N}\}$ *are total families for* $L^p(\Omega)$ *in the sense of* (1.3) *for all* $p \in [1,\infty]$, (b) *there exists* $\epsilon_0 > 0$ *such that the sets* $\{e_{j,k}^{(i)};\ i = 2,4,\ j,k \in \mathbb{N}\}$ *are total families for* $L^p(\Omega)$ *in the sense of* (1.3) *for all* $p \in (1,\infty]$ *if* $0 < \epsilon \leq \epsilon_0$.

We do not obtain a result in part (b) for $p = 1$ because the proof employs a property of the mollifiers that is valid in $L^q(\Omega)$ when $1 \leq q < \infty$ but not when $q = \infty$.

We now discuss our results that functions integrable in the general Denjoy sense belong to F_p spaces. Since the requirements for a function to be integrable in these nonabsolute integration theories are more stringent in higher dimensions, we discuss the cases $n = 1$ and $n \geq 2$ separately.

PROPOSITION 3.2. *Consider the sets* $\{e_\gamma^{(i)},\ \gamma \in \Gamma,\ i = 1,2$ *(or* $i = 3,4)\}$ *with* $n = 1$ *and denote by* $F_p^{(i)}$ *the corresponding* F_p *spaces. Suppose that* $a_\infty^{(i)} < \infty$, $i = 1,\ldots, 4$. *If* f *is integrable in the general Denjoy sense, then* $f \in F_p^{(1,3)}$ *for all* $p \in [1,\infty]$ *and* $f \in F_p^{(2,4)}$ *for all* $p \in (1,\infty]$.

In the proofs one sets $\Omega = [a,b]$, where the interval may be of either finite or infinite length. The proofs rely on the facts that the following seminorm can be defined on the set of general Denjoy integrable functions on $[a,b]$ [1,10,11]:

$$\|f\|_{D_1} = \sup_{x\in[a,b]} \left| \int_a^x f(y)dy \right|, \tag{3.3}$$

where the integral is interpreted in the general Denjoy sense; and that

$$\left|\int_a^b f(x)g(x)dx\right| \leq \|f\|_{D_1}(|g(b)| + V(g;\ [a,b])), \tag{3.4}$$

where the integral is defined as in (3.3) and g is of bounded variation on $[a,b]$, $V(g; [a,b])$ denoting its total variation over $[a,b]$.

When $n \geq 2$ there are many nonequivalent integrals that are more general than the Riemann and Lebesgue integrals [1,2,12]. For our purposes, the most useful is the one defined in [1] (which we have generalized from dimension $n = 2$ to $n \geq 2$). In the sequel we will refer to these integrals as *CD-integrals*.

PROPOSITION 3.3. *Consider the sets* $\{e_\gamma^{(i)}, \gamma \in \Gamma, i = 3,4\}$ *in dimensions* $n \geq 2$ *and denote the corresponding* F_p *spaces by* $F_p^{(i)}$, $i = 3,4$. *If* f *is CD-integrable, then* $f \in F_p^{(3)}(\Omega)$ *for all* $p \in [1,\infty]$ *and* $f \in F_p^{(4)}(\Omega)$ *for all* $p \in (1,\infty]$.

In the proofs one denotes by $\Omega = \prod_{i=1}^{n} [a_i,b_i]$ a finite product of compact intervals in $\mathbb{R}^n$, $n \geq 2$. The seminorms $\|f\|_{D_n} = \sup_{(x_1,\dots, x_n)\in I(\underset{\sim}{x})} |\bar{F}(x_1,\dots, x_n)|$, which generalize the seminorms (3.3), are finite for CD-integrable functions provided that the primitive function $\bar{F}$ satisfies a certain Hölder-type condition on each perfect subinterval of each "component interval" $[a_i,b_i]$ $(i = 1,\dots, n)$ of Ω where $I(\underset{\sim}{x}) \equiv \prod_{i=1}^{n} [a_i,x_i] \subset \Omega$. If f is CD-integrable (with $\bar{F}$ satisfying the condition mentioned above) and g_i is absolutely continuous on $[a_i,b_i]$, $i = 1,\dots,n$, then one obtains an estimate which generalizes (3.4),

$$(3.5) \qquad \left|\int_I fg_1g_2 \cdots g_n dx_1 \cdots dx_n\right| \leq \|f\|_{D_n} \prod_{i=1}^{n} (|g_i(\beta_i)| + V(g_i; [\alpha_i,\beta_i])),$$

for each "subinterval" $I = \prod_{i=1}^{n} [\alpha_i,\beta_i] \subset \Omega$, where V denotes a one-dimensional total variation.

Of the many Denjoy-type integrals in dimensions $n \geq 2$, we have chosen generalizations of the definition in [1] because for them it is possible to obtain the estimates (3.5) which are analogous to the one-dimensional results (3.4). For a comparison of a number of such integrals in the case n = 2, we refer to [12].

We have the following connection between the $F_p(\Omega)$ spaces and the Schwartz distributions, $D'(\Omega)$.

PROPOSITION 3.4. *Consider the sets* $\{e_\gamma^{(i)},\ \gamma \in \Gamma\}$ *for* $i = 2,4$. *Then* $D'(\Omega) \subset F_p^{(i)}(\Omega)$ *for all* $p \in (1,\infty]$.

We note that the implications in the preceding proposition go in the opposite direction compared to the function spaces usually employed in analysis. Thus, for example, the elements of Sobolev spaces are distributions with some additional restrictions. We now discuss in more detail the relationship between the F_p spaces and Schwartz distributions. Several researchers have investigated the relationship between nonabsolutely integrable functions and Schwartz distributions in the one-dimensional case [13-16]. The following theorem generalizes these results to the case of general Denjoy integrable functions when n = 1 and to CD-integrable functions when $n \geq 2$.

THEOREM 3.5. *Let*

$$T_f(\varphi) = \int_{\mathbb{R}^n} f(x)\varphi(x)dx, \quad \varphi \in D(\mathbb{R}^n), \tag{3.6}$$

where f *is locally integrable in the general Denjoy sense when* n = 1 *and in the* CD *sense when* $n \geq 2$, *and the integrals* (3.6) *have the indicated interpretations in these respective cases. Then, in each case,* T_f *is a distribution; i.e., it is a continuous linear functional on* $D(\mathbb{R}^n)$.

Using the results of Theorem 3.5, we can define similar generalized derivatives to the well-known Schwartz distributional derivatives by interpreting the defining integrals in the general Denjoy sense for n = 1 and in the CD sense for $n \geq 2$. For convenience, we will refer to these derivatives as GD-distributional derivatives. Then we can define norms

$\|\cdot\|_{F_p^m}$ ($m \in \mathbb{N}$, $1 \leq p \leq \infty$) formally obtained by replacing the L^p norms in the usual definitions of the Sobolev norms by F_p norms and by interpreting the derivatives as GD distributional derivatives. If we then attempt to define analogous spaces to the Sobolev spaces by using the norms $\|\cdot\|_{F_p^m}$, it is not necessary to make the separate assumptions that $f \in F_p$ because this follows from Propositions 3.2 and 3.3 (for the choices of $\{e_\gamma, \gamma \in \Gamma\}$ therein indicated). However, there is a complication in this procedure which is not present in the definition of Sobolev spaces due to the fact that elements of F_p need not be GD-integrable, as the following example shows.

EXAMPLE 3.6. It is well-known in integration theory that the series $f(x) = \sum_{m=2}^{\infty} \frac{\sin mx}{\log m}$ is a well-defined 2π-periodic function on $\mathbb{R}$, but is not GD-integrable. However, it can be shown that $f \in F_p(0,2\pi)$ for all $p \in [1,\infty)$ by choosing $\{e_m\}$ as the following complete orthonormal set on $L^2(0,2\pi)$:

$$(3.7) \qquad e_m(x) = (2\pi)^{-1/2} \exp(imx), \ m \in \mathbb{Z},$$

and choosing $\{t_r\}$ by

$$(3.8) \qquad t_r = \begin{cases} 0, & r = 0 \\ \frac{3}{\pi^2} r^{-2}, & r \neq 0. \end{cases}$$

If $1 < p < \infty$, there is a method by which Sobolev spaces can be embedded in F_p-type spaces similarly to the way that L^p spaces are embedded in F_p spaces which bypasses the difficulties mentioned above concerning the GD distributional derivatives. The idea is based on the fact that, if $1 < p < \infty$, the Sobolev spaces $W_p^m(\mathbb{R}^n)$ coincide with

$$X_p^m(\mathbb{R}^n) \equiv \{g_\alpha * u = (-\Delta)^{-\alpha/2} u;\ u \in L^p(\mathbb{R}^n),$$

$$0 < \alpha \leq m,\ 0 < \alpha < n\},$$

where g_α denote the Riesz potentials ([17], p. 66). Since $g_\alpha \in L^1(\mathbb{R}^n)$, it follows from Young's inequality for convolutions that $g_\alpha * u \in L^p(\mathbb{R}^n)$ when $u \in L^p(\mathbb{R}^n)$ and $0 < \alpha < n$. Thus, if we define $\mathfrak{F}_p^m(\mathbb{R}^n)$ to be the respective completions of $X_p^m(\mathbb{R}^n)$ in $F_p(\mathbb{R}^n)$, we see that $W_p^m(\mathbb{R}^n)$ are embedded in $F_p(\mathbb{R}^n)$ when $1 < p < \infty$.

We conclude this section with an example which shows that the inclusions stated in Proposition 3.4 are proper when $n = 1$. We conjecture that this conclusion remains valid when $n \geq 2$.

<u>EXAMPLE 3.7</u>. On $\mathbb{R}$, the function

$$\sum_{m=1}^{\infty} \sin mx = \frac{\sin x}{2(1 - \cos x)} \tag{3.9}$$

is not locally integrable in any neighborhood of the origin and therefore does not define a Schwartz distribution. On the other hand, the function (3.9) belongs to $F_p(0,2\pi)$ for all $p \in [1,\infty)$ if we again choose the sets $\{e_m\}$, $\{t_r\}$ as in (3.7) and (3.8), respectively. The proof is similar to the proof of Example 3.6.

4. <u>RELATIONS BETWEEN DIFFERENT</u> F_p <u>SPACES</u>.

Thus far we have defined F_p spaces, described some of their properties, and have noted some possible choices for the associated family of functions $\{e_\gamma, \gamma \in \Gamma\}$. However, we have not yet discussed conditions on either the numbers $\{t_\gamma, \gamma \in \Gamma\}$ or the functions $\{e_\gamma, \gamma \in \Gamma\}$ for which the corresponding F_p spaces are equivalent or inequivalent. We will discuss results of these types in the present section.

<u>PROPOSITION 4.1</u>. *Let* $\Omega \subset \mathbb{R}^n (n \geq 1)$ *be open and suppose that* A *is an unbounded symmetric operator on* $C_0^\infty(\Omega) \subset L^2(\Omega)$ *with a selfadjoint extension* $\tilde{A}$ *which has a purely discrete spectrum* $\{\lambda_j\}_{j=-\infty}^{\infty}$ *with finite multiplicity with corresponding eigenfunctions* $\{e_j\}_{j=-\infty}^{\infty}$. *Then, for each* $p \in [1,\infty)$, *the spaces* $\mathfrak{H}_p(\Omega)$ *with norm*

$$\|\cdot\|_{\tilde{H}_p(\Omega)} = \left(\sum_{j=-\infty}^{\infty} t_j\ |(\tilde{A}\ \cdot, e_j)|^p\right)^{1/p}$$

and $S_p(\Omega)$ *with norm*

$$\|\cdot\|_{S_p(\Omega)} = \left(\sum_{j=-\infty}^{\infty} t_j'|(\cdot, e_j))|^p\right)^{1/p}$$

are equivalent in the sense that the sequences of positive numbers $\{t_j\}_{j=-\infty}^{\infty}$, $\{t_j'\}_{j=-\infty}$ *can be chosen such that* $\sum_{j=-\infty}^{\infty} t_j = \sum_{j=-\infty}^{\infty} t_j' = 1$ *and* $\sum_{j=-\infty}^{\infty} t_j\ |\lambda_j|^p < \infty$ *so that* $\|\psi\|_{\tilde{H}_p(\Omega)} = \left(\sum_{j=-\infty}^{\infty} t_j|\lambda|^p\right)^{1/p} \|\psi\|_{S_p(\Omega)}$ *for all* $\psi \in \tilde{H}_p(\Omega) \cap S_p(\Omega)$.

The class of operators considered in ([18], Theorem 14.6) satisfy the conditions of this proposition. Proposition 4.1 shows that differential operators may be "removed" from the definition of the norms of some F_p-type spaces. More information can be obtained about the equivalence of F_2 spaces if one considers the case in which the functions $\{e_\gamma,\ \gamma \in \Gamma\}$ form complete orthonormal sets of basis functions for the corresponding $L^2(\Omega)$ spaces.

<u>PROPOSITION 4.2</u>. *Consider two spaces* $F_2^{(a,b)}(\Omega)$ *with norms of the form* (1.4) (*with* $p = 2$) *with corresponding numbers* $\{t_\gamma^{(a,b)},\ \gamma \in \Gamma^{(a,b)}\}$ *satisfying* (1.2) *and functions* $\{e_\gamma^{(a,b)},\ \gamma \in \Gamma^{(a,b)}\}$ *which are each complete orthonormal bases for* $L^2(\Omega)$. *Then, the* $F_2^{(a,b)}(\Omega)$ *norms are equivalent if there exists a constant* $C > 0$ *such that*

$$(4.1)\qquad \left|\sum_{j\in\Gamma^{(a)}} t_j^{(a)}(e_j^{(a)}, e_k^{(b)})(e_{k'}^{(b)}, e_j^{(a)})\right| \le C^2 t_k^{(b)}\ \delta_{kk'}$$

for all $k \in \Gamma^{(b)}$ *and a constant* $B > 0$ *such that*

$$(4.2)\qquad \left|\sum_{k\in\Gamma^{(b)}} t_k^{(b)}(e_k^{(b)}, e_j^{(a)})(e_{j'}^{(a)}, e_k^{(b)})\right| \le B^2\ t_j^{(a)}\ \delta_{jj'}$$

for all $j \in \Gamma^{(a)}$.

We note that the summation inside the absolute value sign in (4.1) [(4.2)] reduces to $\delta_{jj'}[\delta_{kk'}]$ when $t_j^{(a)} = 1[t_k^{(b)} = 1]$ for all $j \in \Gamma^{(a)}$ $[k \in \Gamma^{(b)}]$. However, when $\{t_j^{(a)}, j \in \Gamma^{(a)}\}$ $[\{t_k^{(b)}, k \in \Gamma^{(b)}\}]$ satisfies (1.2), one does not expect that the orthogonality of the quantities on the left-hand sides of (4.1) and (4.2) will be preserved. The following example illustrates this statement.

EXAMPLE 4.3. Consider the space $L^2(-1,1)$ and the two complete orthonormal sets:

$$e_m^{(a)}(x) = \frac{1}{\sqrt{2}} \exp(i\pi m x), \ m \in \mathbb{Z},$$
$$e_m^{(b)}(x) = \sqrt{\frac{2m+1}{2}}\, P_m(x), \ m = 0,1,\ldots,$$

where $\{P_m(x);\ m = 0,1,\ldots\}$ are the Legendre polynomials. Then ([19], p. 231),

$$\overline{a_{jk}} \equiv \int_{-1}^{1} \overline{e_j^{(a)}(x)}\, e_k^{(b)}(x)dx = \sqrt{\frac{2k+1}{2}}\, i^{-k} \frac{J_{k+1/2}(\pi j)}{j^{1/2}} ,$$

where $J_{k+1/2}$ is a Bessel function of the first kind. Then, if we choose

$$t_j^{(a)} = \begin{cases} 0, & j = 0 \\ \frac{3}{\pi^2} |j|^{-2}, & j \neq 0, \end{cases}$$

we find, for example, that

$$\sum_{j=-\infty}^{\infty} t_j^{(a)}\, \overline{a_{j2}}\, a_{j4} = -\frac{1}{7\sqrt{5}} \neq 0.$$

ACKNOWLEDGMENTS

This research was supported under NSF grant DMS-8813313, AFOSR contract F49620-89-C-0079 and ARO contract DAAL03-89-C-0038.

REFERENCES

1. V.G. Čelidze and A.G. Džvaršeišvili, *The Theory of the Denjoy Integral and Some Applications*, World Scientific, Singapore (1989).
2. R. Henstock, *The General Theory of Integration*, Clarendon Press, Oxford (1991).
3. S. Saks, *Theory of the Integral*, Monografie Matematyezne, Warsaw (1937).
4. L.C. Evans, *Weak Convergence Methods for Nonlinear Partial Differential Equations*, Conference Board of the Mathematical Sciences, no. 74, Amer. Math. Soc., Providence, R.I. (1990).
5. T.L. Gill, *Trans. Amer. Math. Soc.*, 279, 617-634 (1983).
6. T.L. Gill and W.W. Zachary, *J. Mathematical Phys.*, 28, 1459-1470 (1987).
7. J. Kuelbs, *J. Functional Analysis*, 5, 354-367 (1970).
8. V.Y. Steadman, *Theory of Operators on Banach Spaces*, Ph.D. thesis, Howard University (1988).
9. Y. Meyer, *Ondelettes et Opérateurs I. Ondelettes*, Hermann, Paris (1990).
10. A. Alexiewicz, *Colloq. Math.*, 1, 289-293 (1948).
11. W.L.C. Sargent, *J. London Math. Soc.*, 28, 438-451 (1953).
12. K.M. Ostaszewski, *Henstock Integration in the Plane*, *Memoirs Amer. Math. Soc.*, 63, No. 353 (1986).
13. S. Foglio, *Proc. London Math. Soc.*, (3), 18, 337-348 (1968).
14. S.F.L. de Foglio, *J. London Math. Soc.*, (2), 2, 14-18 (1970).
15. S.F.L. de Foglio and R. Henstock, *J. London Math. Soc.*, (2), 6, 693-700 (1973).
16. P.Y. Lee, *Amer. Math. Monthly*, 77, 984-987 (1970).
17. W.P. Ziemer, *Weakly Differentiable Functions*, Springer-Verlag, New York (1989).
18. S. Agmon, *Lectures on Elliptic Boundary Value Problems*, D. van Nostrand, Princeton, New Jersey (1965).
19. W. Magnus, F. Oberhettinger, and R.P. Soni, *Formulas and Theorems for the Special Functions of Mathematical Physics*, Springer-Verlag, New York (1966).

Third International Colloquium on Differential Equations pp. 119-132 (1993)
D. Bainov and V. Covachev (Eds)

Global existence of holomorphic solutions of differential equations with complex parameters

Joji KAJIWARA
Department of Mathematics, Faculty of Science, Kyushu University 33, Fukuoka 812, Japan

Abstract. Let D be a cylindrical domain in $\mathbf{C}^n$, M be a Stein manifold and T be a linear partial differential operator concerning variables $z \in \mathbf{C}^n$ with coefficients holomorphic in $(z,r) \in D \times M$. We give a neceessary and sufficient condition that , for any holomorphic function g in $D \times M$, the equation $Tf = g$ has a global holomorphic solution f in $D \times M$. The condition concerns with topological properties of D and analytic properties of T.

Keywords: Global existence, Holomorphic solutions

Introduction

Early in 1956 L. Ehrenpreis [1] discoursed on an application of the sheaf theory to differential equations and gave a criterion for the existence of global solutions of differential equations $Tf = g$ in a domain D when the local existence of solutions for g are assured. J. Kajiwara [2] applied this Ehrenpreis' method to linear ordinary differential equations with meromorphic coefficients and gave a necessry and sufficient condition for the global existence in the meromorphic category. In the holomorphic category the condition is that D is either a simply connected domain or a doubly connected domain without non trivial global single-valued holomorphic homogeneous solutions in D.

H. Suzuki [11] stated that the golbal existence of differential eqution $\partial u/\partial x_1 = f$ with complex parameter depends on the Steinness of the sets of cuts over the parameter space. J . Kajiwara-Y. Mori [7] connected [2] with [11] and discussed the global existence of more general ordinary differential equations with complex parameters. J.Kajiwara-K.H.Shon[8] generalized the results to the case of parameter spaces Stein.

Concerning partial differential equation, J. Kajiwara [3]-[6] discussed several concrete cases. In the paper [10] K. H. Shon reported promptly the main theorem and showed sheaf-theoretically a route of its proof. The aim of the lecture in this third Colloquium is to give a necessary and sufficient condition for the global existence stated in the last paragraph in case of polycylindrical domains. This is a preparation to the fourth Colloquium where we will discuss the case that the cuts are cylindrical but vary with the parameter $r \in M$.

1. Global existence and local existence

In this report, a *connected and open* set in a topological space is called a *domain*. Let m be a positive integer and, for each integer j with $1 \leq j \leq m$, D_j be a domain in the complex plane $\mathbf{C}$. We put

$$D = D_1 \times D_2 \times \cdots \times D_m \qquad (1.1).$$

The set D is a domain of holomorphy in the complex m-space $\mathbf{C}^m$. Let M be a Stein manifold. We use the manifold M as a parameter space, denote by r a point of M and regard it as a complex parameter. We consider the product manifold $D \times M$. For each i with $1 \leq i \leq m$, let $a^i = (a^i_{jk}(z, r))$ be a square matrix of degree m whose each (j, k) element $a^i_{jk}(z, r)$ is holomorphic function in $D \times M$. For each i with $1 \leq i \leq m$, let T_i be a differential operator defined by

$$T_i = \frac{\partial}{\partial z_i} + a^i \qquad (1.2).$$

Let $O_{D \times M}$ be the sheaf of germs of all holomorphic functions on $D \times M$. Then each differential operator T_i defines the sheaf homomorphism

$$T_i : (O_{D \times M})^m \longrightarrow (O_{D \times M})^m \qquad (1.3).$$

of $(O_{D \times M})^m$.

We put

$$S = T_1 T_2 \cdots T_m \tag{1.4}$$

and, for any positive integer i with $1 \leq i \leq m$, we also put

$$S_i = T_1 T_2 \cdots T_i \qquad (1.5).$$

Let

$$H(D \times M) = H^0(D \times M, (O_{D \times M})^m) \tag{1.6}$$

be the set of global sections of $(O_{D \times M})^m$ over $D \times M$. $H^0(D \times M, O_{D \times M})$ is the **C**-module of all holomorphic functions on $D \times M$. $H(D \times M)$ is the $H^0(D \times M, O_{D \times M})$-module of all m-column vector valued holomorphic functions on $D \times M$.

Let Ker S and Im S be, respectively, the kernel and image of the homomorphism

$$S\colon H(D \times M) \longrightarrow H(D \times M) \qquad (1.7).$$

For any positive integer i with $m \geq i \geq 1$, let Ker S_i and Im S_i be , respectively, the kernel and image of the homomorphism

$$S_i : H(D \times M) \longrightarrow H(D \times M) \qquad (1.8).$$

Their elements are m-column vector valued global holomorphic functions on $D \times M$.

We obtain the following lemma by induction with on i ($1 \leq i \leq m$):

PROPOSITION 1_i. *If D_1 , D_2, ... and D_i are simply connected, then for the above S_i , we have*

$$H(D \times M) = S_i \; (H(D \times M)) \qquad (1.9).$$

For any p with $1 \leq p \leq m$, we consider a short exact sequence of sheaves

$$0 \longrightarrow \mathrm{Ker} S_p \longrightarrow (O_{D \times M})^m \xrightarrow{S_p} \mathrm{Im}\, S_p \longrightarrow 0 \tag{1.10}$$

where Ker $S_p \longrightarrow (O_{D \times M})^m$ is the canonical injection. The above short exact sequence of sheaves induces the following long exact sequence of cohomologies over $D \times M$:

$$0 \longrightarrow H^0(D \times M, \operatorname{Ker} S_p) \longrightarrow H(D \times M) \xrightarrow{S_p}$$

$$H^0(D \times M, \operatorname{Im} S_p) \longrightarrow H^1(D \times M, \operatorname{Ker} S_p) \longrightarrow$$

$$H^1(D \times M, (O_{D \times M})^m) \longrightarrow \cdots \qquad (1.11)$$

Since $D \times M$ is a domain of holomorphy from the theorem of Weierstrass, we have

$$H^1(D \times M, (O_{D \times M})^m) = 0 \qquad (1.12)$$

from the theorem of Oka-Cartan-Serre. From Proposition 1, we have $\operatorname{Im} S_p = (O_{D \times M})^m$. Hence we have the following Proposition:

PROPOSITION 2. *For the above D, M and* S_p, *there holds*

$$H^1(D \times M, \operatorname{Ker} S_p) = H(D \times M) / S_p(H(D \times M)) \qquad (1.13)$$

i.e.,

$$H^1(D \times M, \operatorname{Ker} S_p) = 0 \qquad (1.14)$$

if and only if

$$H(D \times M) = S_p(H(D \times M) \qquad (1.15).$$

Especially, in case that $p = m$, we have the following proposition:

PROPOSITION 3. *For the domain* $D \times M$ *and the differential operator S given in* (1.1)-(1.7), *there holds*

$$H^1(D \times M, \operatorname{Ker} S) = H(D \times M) / S(H(D \times M)) \qquad (1.16),$$

i.e.,

$$H^1(D \times M, \operatorname{Ker} S) = 0 \qquad (1.17)$$

if and only if

$$H(D \times M) = S(H(D \times M)) \qquad (1.18).$$

By induction with respect to p, we have the following Proposition:

PROPOSITION 4. *If, for an integer p with* $1 \leq p \leq m$, *there holds*

$$H^1(D \times M, \operatorname{Ker} S_p) = 0 \qquad (1.19),$$

then we have

$$H^1(D \times M, \operatorname{Ker} T_1) = 0 \qquad (1.20).$$

PROPOSITION 5. *Let p be a positive integer with* $1 \leq p \leq m$. *For the domain D, the Stein manifold M and the operator* S_p, *if* D_1, D_2, ... *and* D_p *are simply connected, then we have*

$$H^1(D \times M, \text{Ker } S_p) = 0 \qquad (1.21).$$

2. Analytic prolongation round a closed curve

Let i be a positive integer with $1 \leq i \leq m$. We seperate the i-th coordinate z_i of a point $z = (z_1, z_2, z_3, \ldots, z_m)$ from the others, adopt notations $z_i' = z_i$, $z_i'' = (z_1, z_2, \ldots, z_{i-1}, z_{i+1}, \ldots, z_m)$ and $D_i' = D_i$, $D_i'' = D_1 \times D_2 \times \cdots \times D_{i-1} \times D_{i+1} \times \cdots \times D_m$ and identify the point z of D with the point (z_i', z_i'') of $D_i' \times D_i''$.

Let X_i be a simply connected subdomain of D_i which is not neccessarily simply connected. Let z_i^0 be any point of X_i and $f(z_i'', r) = (f_1, f_2, \ldots, f_m)$ be any element of H($D_i'' \times M$). The problem

$$T_i u = 0, \quad u(z_i^0, z_i'', r) = f(z_i'', r) \qquad (2.1).$$

has a unique single-valued holomorphic solution in $X_i \times D_i'' \times M$, which is the solution $u = (u_1, u_2, \ldots, u_m)$ of the integral equation

$$u_j(z_i', z_i'', r) = f_j(z_i'', r) - \int_{z_i^0}^{z_i} \left(\sum_{k=1}^{m} a^i_{jk}(s_i, z_i'', r) u_k(s_i, z_i'', r)\right) d s_i \quad (1 \leq j \leq m) \qquad (2.2),$$

which is obtained by the method of successive approximations and which is denoted by

$$M^i(z_i; z_i^0; z_i'', r; X_i) = \exp\left(-\int_{z_0^i}^{z_i} a^i(s_i, z_i'', r)\, d s_i ; X_i\right) \qquad (2.3).$$

$$u(z_i', z_i'', r) = M^i(z_i; z_i^0; z_i'', r; X_i)\, f(z_i'', r) \qquad (2.4)$$

usnig the symbol exp formally for the sake of our memory. For any point z_i fixed in X_i, $M^i(z_i; z_i^0; z_i'', r; X_i)$ is a linear mapping of the linear space $H(D_i'' \times M)$ regarded as the linear space of initial value into the same $H(D_i'' \times M)$ regarded as the linear space of value $u(z_i', z_i'', r)$ at z_i of solutions to the homogeneous eqution $T_i u = 0$. By the uniqueness of the initial value problems, the mapping (2.4) is bijective and the matrix (2.3) is regular.

Let z_i^1 be any point of the simply connected subdomain X_i. For u defined by (2.1), $u(z_i^1, z_i'', r)$ is its value at the point z_i^1. we have the matrix relation

$$\mathrm{M}^i(z_i; z_i^0; X_i) = \mathrm{M}^i(z_i; z_i^1; X_i)\, \mathrm{M}^i(z_i^1, z_i^0; X_i) \qquad (2.5).$$

So we can regard M^i as non commutative exponentials.

Now, we consider the case that the i-th domain D_i is not simply connected. We take two simply connected domains D_i d and D_i g which cover D_i and denote their intersection by E_i . Take not that the characters d , g, b, h stand for left = gauch, bottom = bas, top = haut in French. We take a starting point z_i^{b} in the simply connected domain D_i d. Let

$$K_i = \{\ z_i = k_i\ (t)\ ;\ 0 \leq t \leq 1\ \} \qquad (2.6)$$

be a closed Jordan curve in D_i with counter-clock wise orientation and with z_i^{b} as the starting and end point. We assume that the part

$$K_i\,\mathrm{d} = \{\ z_i = k_i\ (t)\ ;\ 0 \leq t \leq 1/2\ \} \qquad (2.7)$$

is contained in D_id, but the whole K_i is not contained in X_i , the part

$$K_i\,\mathrm{g} = \{\ z_i = k_i\ (t)\ ;\ 1/2 \leq t \leq 1\ \} \qquad (2.8)$$

is contained in D_i g,the relatively compact connected component of D_i - K_i is not contained in D_i and K_i is one element of a homology base of D_i .

Let $f(z_i'', r)$ be any element of $\mathrm{H}(D_i'' \times M)$. For any $(z_i'', r) \in D_i'' \times M$, we consider an initial value problem

$$T_i\, u = 0,\ u(z_i,\ z_i'',\ r) = f(z_i'',\ r) \qquad \text{for } z_i = z_i^{\mathrm{b}} \qquad (2.9)$$

in D_i d and continue it analytically along the closed curve K_i . After analytic prolongation of the solution $u = (u_1, u_2, \dots, u_m)$ to the initial problem (2.9) along the first half K_i d of the closed curve K_i , we arrive at a relay point

$$z_i^{\mathrm{h}} = k_i\ (1/2) \in E_i \qquad (2.10).$$

By the definition (2.4), its value at z_i^{h} is represented by

$$u(z_i^{\mathrm{h}},\ z_i'',\ r) = \mathrm{M}^i(z_i^{\mathrm{h}}; z_i^{\mathrm{b}}\ ; D_i\,\mathrm{d})\ f(z_i'',\ r) \qquad (2.11).$$

We consider another initial value problem

$$T_i\, u = 0\ ,\ u(z_i^{\mathrm{h}}\ ,\ z_i'',\ r) = \mathrm{M}^i(z_i^{\mathrm{h}}; z_i^{\mathrm{b}}\ ; D_i\,\mathrm{d})\ f(z_i'',\ r) \qquad (2.12)$$

in the simply connected domain D_ig . The continuation of (2.12) along K_i g is regarded as a succession of the continuation of (2.9) along K_i d. By the uniqueness of initial value problems and (2.4), the value of u given by (2.12) is the right hand-side of

$$\mathrm{Monod}(K_i\ ; z_i^{\mathrm{b}}; z_i'',\ r)\ f(z_i'',\ r) =$$

$M^i(z_i^b; z_i^h; D_i\ g)\ M^i(z_i^h; z_i^b; D_i\ d)\ f(z_i\ '', r)$ (2.13)
and is denoted by the left hand-side of (2.13). Then the matrix
$\text{Monod}(K_i; z_i^b; z_i\ '', r) = M^i(z_i^b; z_i^h; D_i\ g)\ M^i(z_i^h; z_i^b; D_i\ d)$ (2.14)
denote the result of analytic prolongation round the closed curve K_i in D_i . Since the two matrices in the right hand-side of (2.14) are regular as we show it above, the matrix $\text{Monod}(K_i; z_i^b; z_i\ '', r)$ is also regular for any $(z_i\ '', r)$ of $D_i'' \times M$.

This $\text{Monod}(K_i; z_i^b; z_i\ '', r)$ is regarded as a $m \times m$ matrix whose elements belong to $H(D_i'' \times M)$ and invertible. We put

$L^i(K_i; z_i^b; z_i\ '', r) = \text{Monod}(K_i; z_i^b; z_i\ '', r) - 1$ (2.15)
where 1 denotes the $m \times m$ unit matrix. Then the $m \times m$ matrix (2.15) explains the fluctuation of values of the initial value problem (2.9) after the analytic prolongation along the closed Jordan curve K_i .

3. Connectivity of D_i under the assumption $H^1(D \times M, \text{Ker}\ T_i) = 0$

PROPOSITION 6. *Under the notation of the preceding paragraph, if there holds*

$$H(D \times M) = T_i\,(H(D \times M)) \quad (3.1)$$

that is,

$$H^1(D \times M, \text{Ker}\ T_i) = 0 \quad (3.2),$$

then either D_i is a simply connected domain or D_i is a doubly connected domain with regular matrix $L^i(K_i; z_i^b; z_i\ '', r)$ for a homology base K_i .

Proof. Assume that D_i were neither simply nor doubly connected. Then there exist two closed Jordan curves Kb and Kh ;

$$Kq = \{ z_i = kq(t) ;\ 0 \leq t \leq 1 \}\ \ (q=b, h) \quad (3.3)$$

and simpy connected subdomains D_i d and D_i g of D_i satisfying the following conditions:

(a) $\{D_i\ d, D_i\ g\}$ is an open covering of D_i .

(b) The orientations of Kb and Kh with increasing t are counter-clock wise. Kb and Kh belong to a homology base of D_i .

$kh(1/2) = kb(1/2)$. $Khg = \{kh(t);\ 0 \leq t \leq 1/2\}$ and $Kbd = \{kb(t);\ 0 \leq t \leq 1/2\}$ are contained , respectively, in D_i g and D_i d.

$Khd = \{kh(t);\ 1/2 \leq t \leq 1\}$ and $Kbg = \{kb(t);\ 1/2 \leq t \leq 1\}$ are contained , respectively, in D_i d and D_i g.

(c) The connected components of the intersection E_i of D_i d and D_i g consist of simply connected components of an open subset E_i 0, simply connected subdomains E_i h and E_i b . The open set E_i 0 contains $Kh(0) = kh(1)$, the simply connected domain E_ih contains

$$z_i{}^{h} = kh(1/2) = kb(1/2) \qquad (3.4),$$

and the simply connected domain E_2b contains

$$z_i{}^{b} = kb(0) = kb(1) \qquad (3.5).$$

We put

$$Dd = D_i\text{'d} \times D_i\text{''},$$
$$Dg = D_i\text{'g} \times D_i\text{''} \qquad (3.6)$$

and

$$E = E_i \times D_i\text{''},$$
$$Eh = E_i\text{h} \times D_i\text{''},$$
$$Eb = E_i\text{b} \times D_i\text{''},$$
$$E0 = E_i 0 \times D_i\text{''} \qquad (3.7).$$

Then U = {*D*d x *M*, *D*g x *M*} is an open covering of *D* x *M*. By (3.2), we have

$$H^1(U, \text{Ker } T_i) = 0 \qquad (3.8).$$

For any element *h* of H^0(*E*b x *M*, Ker T_i), we attach the element D_i of H^0(*E*b x *M*, Ker T_i) to *E*b x *M*, the element 0 of H^0(*E*h x *M*, Ker T_i) to *E*h x *M*, the element 0 of H^0(*E*0 x *M*, Ker T_i) to *E*0 x *M* and regard this as a cocycle belonging to Z^1(U , Ker T_i). By (3.6), this is a coboundary ,i.e., there exist *h*d of H^0(*E*d x *M*, Ker T_i) and *h*g of H^0(*E*g x *M*, Ker T_i) such that

$$hg - hd = h \qquad (3.9)$$

on Eb x M and

$$hg = hd \qquad (3.10)$$

on *E*h x *M* and *E*0 x *M*. For any fixed point (z_i ", *r*) of D_i " x *M*, starting from the point $z_i{}^{b}$ in E_i b, hd can be analytically continuous along the closed curve K_i as we have discussed in Paragraph 2. Moreover, (3.8) claims that *h*g is a successor at relay point $z_i{}^{h}$ to the result of this analytic continuation and (3.7) claims that the difference coincides with the just given *h*.

Now, we take any element $g(z_i '', r)$ and $f(z_i '', r)$ of $H(D_i '' \times M)$ adopt the solution in $Dd \times M$ of the initial value problem (2.18) as h and adopt the solution in $Dd \times M$ of the problem of the same differential equation $T_i\ u = 0$ with the initial data $f(z_i '', r)$ as hd. Then, substituting $z_i = z_i{}^b$ in (3.9) and taking into account of the definition (2.26), we have

$$L^i(K_i b; z_i{}^b; z_i '', r)\ f(z_i '', r) = g(z_i '', r) \qquad (3.11).$$

Since $g(z_i '', r)$ is arbitrary, the linear transformation $L^i(K_i b; z_i{}^b; z_i '', r)$ is surjective. Since we are now discussing the finite dimensional matrix $L^i(K_i b; z_i{}^b; z_i '', r)$, it is invertible, i. e., regular matrix.

Thus, if we are given the set $H(D_i '' \times M)$ of all initial datum $g(z_i '', r)$'s at the starting point $z_i{}^b$, then the set of all $f(z_i '', r)$'s determined by (3.9) coincides with the full set $H(D_i '' \times M)$. So we may assume that the initial data $f(z_i '', r)$ at the starting point $z_i{}^b$ is arbitrary. Then the continuation of hd along Kbd supplies us the set of $M^i(z_i{}^h; z_i{}^b; D_i\, d)\ f(z_i '', r)$. Since $M^i(z_i{}^h; z_i{}^b; D_i\, d)$ is surjective, this set coincides with the full set $H(D_i '' \times M)$ too. So we may assume any element hd is given at the point $z_i{}^b$.

Now, we see the other closed curve Kh. Since $hg = hd$ holds on $Eh \times M$ and $E0 \times M$, hg is a direct continuation of hd to $Dg \times M$. This means that any solution u of the homogeneous equation $T_i\ u = 0$ is single valued along the closed curve Kh.

Because the two closed curves Kb and Kh in D_i are logically equal, this is a contradiction.

Thus we have proved D_i is either simply connected or doubly connected with $H^0(D \times M, \mathrm{Ker}\, T_i) = 0$. q.e.d.

4. Neccessary and sufficient condition for $H^1(D \times M, \mathrm{Ker}\ T_i) = 0$

Now we consider the case that the target i-th domain D_i is doubly connected and let $z_i{}^b$ be a point in D_i and

$$K = \{\, z_i = k(t)\ ;\ 0 \le t \le 1 \,\} \qquad (4.1)$$

be a closed Jordan curve in D_i with the starting and end point $z_i{}^b$ which is a homology base of D_i. As in the proof of Proposition 6, we associate to the above closed curve K in D_i two subdomains $D_i d$ and $D_i g$ of the domain D_i satisfying the following conditions:

(a) $\{D_i d\ ,\ D_i g\}$ is an open covering of D_i.

(b) The orientations of K with increasing t are counter-clock wise. K is a homology base of D_i.

(c) The connected components of the intersection of D_id and D_ig consist of two simply connected subdomains E_i b and E_i h. E_i b and E_i h contain, respectively,

$$k(0) = z_i{}^{\mathrm{b}} \tag{4.2}$$

and

$$z_i{}^{\mathrm{h}} = k(1/2) \tag{4.3}.$$

We put

$$Dd = D_i\mathrm{d} \times D_i" ,$$
$$Dg = D_i\mathrm{g} \times D_i" \tag{4.4}$$

and

$$E = E_i \times D_i",$$

$$Eb = E_i\,\mathrm{b} \times D_i",$$
$$Eh = E_i\,\mathrm{h} \times D_i" \tag{4.5}.$$

Then U = $\{Dd \times M, Dg \times M\}$ is an open covering of D X M.

Since $D \times M$ is the union of $Dd \times M$ and $Dg \times M$ and since $E \times M$ is the intersection of $Dd \times M$ and $Dg \times M$, we can adopt the exact Mayer-Vietoris cohomology sequence

$$\mathrm{H}^0(D \times M, \mathrm{Ker}\, T_i) \longrightarrow \mathrm{H}^0(Dd \times M, \mathrm{Ker}\, T_i) \oplus \mathrm{H}^0(Dg \times M, \mathrm{Ker}\, T_i) \longrightarrow \mathrm{H}^0(E \times M, \mathrm{Ker}\, T_i) \longrightarrow \mathrm{H}^1(D \times M, \mathrm{Ker}\, T_i) \longrightarrow \mathrm{H}^1(Dd \times M, \mathrm{Ker}\, T_i) \oplus \mathrm{H}^1(Dg \times M, \mathrm{Ker}\, T_i) \longrightarrow \tag{4.6}$$

where the homomorphism

$$\Xi: \mathrm{H}^0(Dd \times M, \mathrm{Ker}\, T_i) \oplus \mathrm{H}^0(Dg \times M, \mathrm{Ker}\, T_i) \longrightarrow \mathrm{H}^0(E \times M, \mathrm{Ker}\, T_i) \tag{4.7}$$

is the canonical substraction

$$\Xi[hd, hg] = hd - hg ,$$
$$[hd, hg] \in \mathrm{H}^0(Dd \times M, \mathrm{Ker}\, T_i) \oplus \mathrm{H}^0(Dg \times M, \mathrm{Ker}\, T_i) \tag{4.8}.$$

PROPOSITION 7. *Assume that the i-th domain D_i is doubly connected. Then the following conditions are equivalent:*

(a) $\mathrm{H}^0(D \times M, (\mathrm{O}_{D \times M})^m) = T_i(\mathrm{H}^0(D \times M, (\mathrm{O}_{D \times M})^m))$ (4.9).

(b) $\mathrm{H}^1(D \times M, \mathrm{Ker}\, T_i) = 0$ (4.10).

(c) *The homomorphism Ξ is surjective* (4.11).

(d) *The matrix* $L^i(K_i; z_i{}^b; z_i{}'', r)$ *defined by* (2.26) *is invertible at each point* (z, r) *of* $D \times M$ (4.12).

(e) $H^0(D \times M, \mathrm{Ker}\, T_i) = 0$ (4.13).

Proof. We show that (d) implies (c). Assume that $L^i(K_i; z_i{}^b; z_i{}'', r)$ is a regular matrix at any point $(z_i{}'', r)$ of $D_i{}'' \times M$. We take any element $u = [ub, uh] \in H^0(E \times M, \mathrm{Ker}\, T_i)$ and seek $[ud, ug] \in H^0(Dd \times M, \mathrm{Ker}\, T_i) \oplus H^0(Dg \times M, \mathrm{Ker}\, T_i)$ with $\Xi[ud, ug] = u$ on $E \times M$ what is equivalent to the following simultaneous equations

$$ug - ud = ub \tag{4.14}$$

in $Eb \times M$ and

$$ug - ud = uh \tag{4.15}$$

in $Eh \times M$ because E is divided into Eb and Eh. We put

$$g(z_i{}'', r) = (M^i(z_i{}^h; z_i{}^b; D_i d))^{-1}\, uh(z_i{}^h, z_i{}'', r) \tag{4.16}$$

and

$$f(z_i{}'', r) = (L^i(K_i; z_i{}^b; z_i{}'', r))^{-1}(ub(z_i{}^b; z_i{}'', r) - g(z_i{}'', r)) \tag{4.17}$$

in $D_i{}'' \times M$. We put

$$ud(z_i, z_i{}'', r) = M^i(z_i{}^h; z_i{}^b; D_i d)\, (f - g)\, (z_i{}'', r) \tag{4.18}$$

in $Dd \times M$ and

$$ug(z_i, z_i{}'', r) = M^i(z_i; z_i{}^b; D_i g)\, \mathrm{Monod}(K_i; z_i{}^b; z_i{}'', r)\; f(z_i{}'', r) \tag{4.19}$$

in $Dg \times M$. We have

$$ug - ud = M^i(z_i; z_i{}^b; D_i g)\, (L^i(K_i; z_i{}^b; z_i{}'', r)\; f(z_i{}'', r) + g) \tag{4.20}$$

in $Eb \times M$. By (5.21) we have

$$\begin{aligned} ug(z_i, z_i{}'', r) &= M^i(z_i; z_i{}^b; D_i g)\, M^i(z_i{}^b; z_i{}^h; D_i g)\, M^i(z_i{}^h; z_i{}^b; D_i d)\; f(z_i{}'', r) \\ &= M^i(z_i; z_i{}^h; D_i g)\, M^i(z_i{}^h; z_i{}^b; D_i d)\; f(z_i{}'', r) \\ &= M^i(z_i; z_i{}^h; D_i d)\, M^i(z_i{}^h; z_i{}^b; D_i d)\; f(z_i{}'', r) \end{aligned} \tag{4.21}$$

in $Eh \times M$. By (2.5), we have

$$ud(z_i, z_i{}'', r) = M^i(z_i; z_i{}^h; D_i d)\, M^i(z_i{}^h; z_i{}^b; D_i d)\; (f - g)\, (z_i{}'', r) \tag{4.22}$$

in Eh × M. Hence we have

$$ug - ud = M^i(z_i; z_i{}^h; D_i d)\, M^i(z_i{}^h; z_i{}^b; D_i d)\; g(z_i{}'', r) \tag{4.23}$$

in $Eh \times M$. By (4.23), one of the surjectivity condition (4.15) is equivalent to the equality

$$M^i(z_i; z_i{}^h; D_i d)\, M^i(z_i{}^h; z_i{}^b; D_i d)\; g(z_i{}'', r) = uh(z_i, z_i{}'', r) \tag{4.24}$$

in $Eh \times M$. By the uniqueness of the initial value problem, it suffices to get (4.24) at $z_i = z_i{}^{h}$. By (4.20), another surjectivity condition (4.14) is equivalent to the equality

$$M^i(z_i\,;\, z_i{}^{h};\, D_i g)\,(L^i(K_i\,;\, z_i{}^{b};\, z_i{}'',\, r)\;\; f(z_i{}'',\, r) + g) = ub(z_i,\, z_i{}'',\, r) \qquad (4.25)$$

in $Eb \times M$. This is the reason why we put (5.19). Since we have

$$M^i(z_i\,;\, z_i{}^{h};\, D_i d) = M^i(z_i\,;\, z_i{}^{h};\, D_i g) \qquad (4.26)$$

in the simply connected subdomain $E_i h$ contained in both $D_i d$ and $D_i g$, we have

$$ug - ud = M^i(z_i\,;\, z_i{}^{h};\, D_i d) g(z_i{}'',\, r) \qquad (4.27)$$

in $Eh \times M$.

THEOREM1. *The necessary and sufficient condition that* $H(D \times M) = T_1(H(D \times M))$ *is that either* D_1 *is simply connected or* D_1 *is doubly connected and* $H^0(D \times M, \mathrm{Ker}\, T_1) = 0$.

5. Main Theorem.

MAIN THEOREM. *Let m be an integer and, for each j with* $1 \le j \le m$, D_j *be a domain in the complex plane* **C**. *We put*

$$D = D_1 \times D_2 \times \cdots \times D_m.$$

Let M be a Stein manifold. We consider the product manifold D x M. For each i with $1 \le i \le m$, *let* $a^i = (a^i{}_{jk}(z, r))$ *be a square matrix of degree m whose element* $a^i{}_{jk}(z, r)$ *are holomorphic functions in D x M. For each i with* $1 \le i \le m$, *let* T_i *be a differential operator defined by*

$$T_i = \frac{\partial}{\partial z_i} + a^i .$$

Let $O_{D \times M}$ *be the sheaf of germs of all holomorphic function on D. Then each differential operator* T_i *defines a sheaf homomorphism*

$$T_i : (O_{D \times M})^m \to (O_{D \times M})^m .$$

We define

$$S = T_1 T_2 \cdots T_m$$

by putting

$$Su = T_1(T_2 (\cdots T_m (u)))$$

for any u of $(O_{D \times M})^m$.

Then, the necessary and sufficient condition for $H(D\times M) = S(H(D\times M))$ *is that* D_i *is either a simply connected domain or a doubly connected domain in* **C** *with invertible* $L^i(K_i\,;\, z_i'', r)$ *where* K_i *is a homology base of* D_i *and* $L^i(K_i\,;\, z_i'', r)$ *is a matrix defined in* (2.26) *for* $i=1,2, \ldots, m$.

Proof. Although T_p's are not commutative, as is proved in Kajiwara-Shon[9], $H^1(D\times M, \text{Ker } T) = 0$ implies $H^1(D\times M, \text{Ker } S_p) = 0$ which implies $H^1(D\times M, \text{Ker } T_p) = 0$ for any $p\in\{1, 2, \cdots, n\}$. By the above Propositions , we have the main theorem.

For the complete proof in Kajiwara-Shon[9], please contact with the address of the facsimile below:

Prof. Joji KAJIWARA
FAX: 0081-92-632-2737

References

[1] L. Ehrenpreis, *Sheaves and differential equations*, Proc. Amer. Math. Soc., **7**(1956), 1131-1139.

[2] J. Kajiwara, *On an application of L. Ehrenpreis' method to ordinary differential equations*, Kodai Math. Sem. Rep. **15**(1963), 94-105..

[3] J. Kajiwara, *Some systems of partial differential equations in the theory of soil mechanics*, Mem. Fac. Sci. Kyushu Univ., **24**(1970), 147-230.

[4] J. Kajiwara, *Some systems of partial differential equations in complex domains*, Mem. Fac. Sci. Kyushu Univ., **25**(1971), 21-143.

[5] J. Kajiwara, Unique holomorphic solution of a singular partial differential equation with a double pole, Periodica Mathematica Hungarica **5**-1 (1974) , 61-71.

[6] J. Kajiwara, *Solution holomorphes globales des e quations differentielles lineaires a valeurs dans un espace de Hilbert et a parametre complexe*, Jap. J. Math., **2**(1976),91-107.

[7] J. Kajiwara and Y. Mori, *On the existence of global holomorphic solutions of differential equations with complex parameters*, Czechoslovak Math. J., **24**(99) (1974), 444-454.

[8] J. Kajiwara and K. H. Shon, *Localization of global existence of solutions of ordinary differential equations with parameter in Stein spaces*, Mem. Fac. Sci. Kyushu Univ., **38**(1984), 91-120.

[9] J. Kajiwara and K. H. Shon, *Global existence of holomorphic solutions of differential equations with complex parameters*-I (to appear in Mem. Fac. Sci. Kyushu Univ.(1993)).

[10] K. H. Shon, *Global holomorphic solutions of differential equations with a parameter in a Stein manifold*, J. Korean Math. Soc. **27**-1 (1990), 820-822.

[11] H. Suzuki, *On the global existence of holomorphic solutions of* $\partial u/\partial x_1 = f$, Sci. Rep. Tokyo Kyoiku Daigaku, **11**(1972), 253-258.

[12] I. Wakabayashi, *Non existence of holomorphic solutions of* $\partial u/\partial z_1 = f$, Proc. Jap. Acad., **44**(1968), 820-822.

Third International Colloquium on Differential Equations pp. 133-146 (1993)
D. Bainov and V. Covachev (Eds)

ON THE RIEMANN–HILBERT PROBLEM FOR THE ONE–DIMENSIONAL SCHRÖDINGER EQUATION

Martin Klaus
Department of Mathematics, Virginia Tech, Blacksburg, VA 24061

Abstract: We study the existence and uniqueness of solutions of a matrix Riemann-Hilbert problem associated with the one-dimensional Schrödinger equation. The connection between this Riemann-Hilbert problem and the non-uniqueness in the inverse problem is investigated.

Keywords: Riemann-Hilbert problem, 1-D Schrödinger equation, Scattering matrix, Inverse scattering

1. INTRODUCTION

In this article we report on some recent work on a matrix Riemann-Hilbert problem associated with the one-dimensional Schrödinger equation. For the most part, but not completely, the results discussed here are from recent joint work with T.Aktosun and C. van der Mee [1] and will be published elsewhere. We will frequently refer the reader to [1] for details.

We consider the one-dimensional Schrödinger equation

$$\psi''(k,x) + k^2\psi(k,x) = V(x)\,\psi(k,x), \qquad -\infty < x < \infty, \tag{1.1}$$

where $V(x)$ is a real-valued potential in $L_1^1 = \{V \mid \int_{-\infty}^{\infty}(1+|x|)|V(x)|\,dx < \infty\}$. Let $f_l(k,x)$ and $f_r(k,x)$ denote the Jost solutions associated with (1.1) such that $f_l(k,x) \sim e^{ikx}$ as $x \to +\infty$ and $f_r(k,x) \sim e^{-ikx}$ as $x \to -\infty$. Put $\psi_l(k,x) = T(k)\,f_l(k,x)$ and $\psi_r(k,x) = T(k)\,f_r(k,x)$. Then the following asymptotic relations hold

$$\psi_l(k,x) = \begin{cases} T(k)e^{ikx} + o(1), & x \to +\infty \\ e^{ikx} + L(k)e^{-ikx} + o(1), & x \to -\infty, \end{cases} \tag{1.2}$$

$$\psi_r(k,x) = \begin{cases} e^{-ikx} + R(k)e^{ikx} + o(1), & x \to +\infty \\ T(k)e^{-ikx} + o(1). & x \to -\infty. \end{cases} \tag{1.3}$$

Here $T(k)$ is the transmission coefficient and $L(k)$ and $R(k)$ are the reflection coefficients from the left and from the right, respectively. For $k \in \mathbb{R}$ the pairs $\{\psi_l(k,x), \psi_r(k,x)\}$ and $\{\psi_l(-k,x), \psi_r(-k,x)\}$ form linearly independent sets of solutions of (1.1). Furthermore.

$\psi_l(-k,x) = \overline{\psi_l(k,x)}$ and $\psi_r(-k,x) = \overline{\psi_r(k,x)}$ when $k \in \mathbb{R}$, where a bar denotes complex conjugation. Let

$$\mathbf{S}(k) = \begin{bmatrix} T(k) & R(k) \\ L(k) & T(k) \end{bmatrix} \tag{1.4}$$

denote the scattering matrix. As a consequence of (1.2) and (1.3), we have that

$$\begin{bmatrix} \psi_r(k,x) \\ \psi_l(k,x) \end{bmatrix} = \mathbf{S}(k) \begin{bmatrix} \psi_l(-k,x) \\ \psi_r(-k,x) \end{bmatrix}, \qquad k \in \mathbb{R}. \tag{1.5}$$

Define $m_l(k,x) = e^{-ikx} f_l(k,x)$, $m_r(k,x) = e^{ikx} f_r(k,x)$, $m(k,x) = \begin{bmatrix} m_l(k,x) \\ m_r(k,x) \end{bmatrix}$ and $\mathbf{J} = \begin{bmatrix} 1 & 0 \\ 0 & -1 \end{bmatrix}$. Then, by (1.1), $m(k,x)$ obeys the vector equation

$$m''(k,x) + 2ik\,\mathbf{J}\,m'(k,x) = V(x)\,m(k,x). \tag{1.6}$$

Letting $\mathbf{q} = \begin{bmatrix} 0 & 1 \\ 1 & 0 \end{bmatrix}$, we see that (1.5) is equivalent to

$$m(-k,x) = \mathbf{G}(k,x)\,\mathbf{q}\,m(k,x), \qquad k \in \mathbb{R}, \tag{1.7}$$

where

$$\mathbf{G}(k,x) = \begin{bmatrix} T(k) & -R(k)e^{2ikx} \\ -L(k)e^{-2ikx} & T(k) \end{bmatrix}. \tag{1.8}$$

If $V \in L_1^1$, then the functions $m_l(k,x)$ and $m_r(k,x)$ are continuous for $k \in \mathbb{R}$ and can be extended analytically in k to $\mathbf{C}^+$. Moreover, $m_l(k,x) \to 1$ and $m_r(k,x) \to 1$ as $k \to \infty$ in $\overline{\mathbf{C}^+}$. Here $\mathbf{C}^+$ denotes the complex upper half-plane and $\overline{\mathbf{C}^+}$ denotes its closure. Similarly, $\mathbf{C}^-$ and $\overline{\mathbf{C}^-}$ denote the complex lower half-plane and its closure, respectively. Thus $m_l(-k,x)$ and $m_r(-k,x)$ can be extended analytically in k to $\mathbf{C}^-$. Hence, when $\mathbf{S}(k)$ is given, (1.7) represents a Riemann-Hilbert problem for the vector function $m(k,x)$. Once (1.7) is solved, the potential can be obtained from (1.6) using

$$V(x) = \frac{m_l''(k,x) + 2ikm_l'(k,x)}{m_l(k,x)} = \frac{m_r''(k,x) - 2ikm_r'(k,x)}{m_r(k,x)}, \tag{1.9}$$

provided, of course, $m_l(k,x)$ and $m_r(k,x)$ both yield the same potential. In this sense, solving the Riemann-Hilbert problem (1.7) amounts to solving the inverse scattering problem for (1.1). In addition to the scattering matrix $\mathbf{S}(k)$ we introduce the matrix

$$\mathbf{J}\,\mathbf{S}(k)\,\mathbf{J} = \begin{bmatrix} T(k) & -R(k) \\ -L(k) & T(k) \end{bmatrix}, \tag{1.10}$$

and consider, in analogy to (1.7), the vector Riemann-Hilbert problem

$$n(-k,x) = \mathbf{J}\,\mathbf{G}(k,x)\,\mathbf{J}\,\mathbf{q}\,n(k,x), \qquad k \in \mathbb{R}. \tag{1.11}$$

It turns out that we must allow for $n(k,x)$ to have a singularity at $k=0$. Thus we seek a solution vector $n(k,x) = \begin{bmatrix} n_l(k,x) \\ n_r(k,x) \end{bmatrix}$ such that, for each fixed $x \in \mathbb{R}$, $n_l(k,x)$ and $n_r(k,x)$ are continuous in k in $\mathbb{R} \setminus \{0\}$ and have analytic continuations to $\mathbf{C}^+$ such that $n_l(k,x) \to 1$ and $n_r(k,x) \to 1$ as $k \to \infty$ in $\overline{\mathbf{C}^+}$. The exact behavior of $n(k,x)$ near $k=0$ depends on the scattering matrix $\mathbf{S}(k)$. Assume for the moment that there exists a potential $U(x)$ corresponding to the scattering matrix $\mathbf{J}\,\mathbf{S}(k)\,\mathbf{J}$. Let $g_l(k,x)$ and $g_r(k,x)$ denote the Jost solutions associated with $U(x)$ and put $n_l(k,x) = e^{-ikx}\,g_l(k,x)$ and $n_r(k,x) = e^{ikx}\,g_r(k,x)$. Similarly to (1.6), we have that

$$n''(k,x) + 2ik\,\mathbf{J}\,n'(k,x) = U(x)\,n(k,x). \tag{1.12}$$

Define the 2×2 matrix $\mathbf{M}(k,x)$ by

$$\mathbf{M}(k,x) = \frac{1}{2}\begin{bmatrix} m_l(k,x) + n_l(k,x) & m_l(k,x) - n_l(k,x) \\ m_r(k,x) - n_r(k,x) & m_r(k,x) + n_r(k,x) \end{bmatrix}. \tag{1.13}$$

Let $\hat{1} = \begin{bmatrix} 1 \\ 1 \end{bmatrix}$ and $\hat{e} = \begin{bmatrix} 1 \\ -1 \end{bmatrix}$. Then

$$m(k,x) = \mathbf{M}(k,x)\hat{1}, \tag{1.14}$$

$$n(k,x) = \mathbf{J}\,\mathbf{M}(k,x)\hat{e}. \tag{1.15}$$

We can combine (1.7) and (1.11) into the matrix Riemann-Hilbert problem

$$\mathbf{M}(-k,x) = \mathbf{G}(k,x)\,\mathbf{q}\,\mathbf{M}(k,x)\,\mathbf{q}, \qquad k \in \mathbb{R}, \tag{1.16}$$

where $\mathbf{M}(k,x)$ is continuous for $k \in \mathbb{R} \setminus \{0\}$ and has an analytic extension in k to $\mathbf{C}^+$, and $\mathbf{M}(k,x) \to \mathbf{I}$ as $k \to \infty$ in $\overline{\mathbf{C}^+}$ for each x. In this paper we only consider transmission coefficients for which the following dichotomy holds: (i) $T(k) = ick + o(1)$ as $k \to 0$ where c is a real nonzero constant, or (ii) $T(k) \to T(0) \neq 0$. Then $R(0) = L(0) = -1$ in case (i) and $0 < |R(0)| = |L(0)| < 1$ in case (ii). For the class L_1^1 this dichotomy is known to be valid **[2]**, **[3]**. We will call a potential $V(x)$ or $U(x)$ generic (exceptional) if $T(k)$ behaves according to (i) ((ii)). Note that $V(x)$ and $U(x)$ have the same transmission coefficients. It will be shown that there always exists a potential $U(x)$ whose scattering matrix is $\mathbf{J}\mathbf{S}(k)\mathbf{J}$, but $U(x)$ will generically be nonunique. In the generic case the potential $U(x)$ cannot lie in L_1^1 since at $k = 0$ the reflection coefficients for $U(x)$ have the 'wrong' value $+1$ instead of -1. Note that (1.16) is a generalization of the standard Riemann-Hilbert problem in the sense that $\mathbf{M}(k,x)$ is not required to be continuous at $k = 0$.

Formulations of the inverse problem as a matrix Riemann-Hilbert problem in the form (1.16), or in different but equivalent forms, have appeared previously in the literature [4],[5], [6]. In [7], Degasperis and Sabatier use a Darboux transformation to construct a one-parameter family of potentials associated with a scattering matrix of the form $\mathbf{J}\mathbf{S}(k)\mathbf{J}$. In some examples such families have also been obtained by different methods [5], [8], [9]. This raises the question of whether the potentials obtained by these methods represent all possible potentials that can be associated with a given scattering matrix of the form $\mathbf{J}\mathbf{S}(k)\mathbf{J}$. Even with respect to the scattering matrix $\mathbf{S}(k)$ one can ask whether there are any potentials outside L_1^1 that have $\mathbf{S}(k)$ as their scattering matrix. We are not able to answer these questions in full generality yet, but we have results if the singularity that $\mathbf{M}(k,x)$ can have at $k = 0$ is suitably restricted. Under these restrictions, we find that there are no potentials associated with $\mathbf{J}\mathbf{S}(k)\mathbf{J}$ other than those given by a one-parameter family, and that there are no potentials outside L_1^1 associated with $\mathbf{S}(k)$.

2. EXISTENCE OF SOLUTIONS

In order to establish the existence of solutions of the Riemann-Hilbert problem (1.16) we need to consider the zero-energy solutions of (1.1), i.e. the solutions of

$$\psi'' + V(x)\psi = 0. \tag{2.1}$$

We divide our analysis into two parts according as V has bound states or not. First suppose that V has no bound states. Two solutions of (2.1) are given by $m_l(0,x)$ and $m_r(0,x)$. It is well known that if V is generic, then $m_l(0,x)$ and $m_r(0,x)$ are linearly independent while if V is exceptional, then $m_l(0,x)$ and $m_r(0,x)$ are linearly dependent. Moreover, $m_l(0,x) = 1 + o(1)$ as $x \to +\infty$, $m_l(0,x) = -c_l\,x + o(x)$ as $x \to -\infty$, $m_r(0,x) = c_r\,x + o(x)$ as $x \to +\infty$ and $m_r(0,x) = 1 + o(x)$ as $x \to -\infty$, where $c_r = c_l > 0$ on account of the Wronskian and the fact that in the absence of bound states $m_l(0,x)$ and $m_r(0,x)$ are strictly positive [10], [1]. In the exceptional case, $c_r = c_l = 0$ and $m_l(0.x)$ and $m_r(0,x)$ are bounded. We refer the reader to [1] for more details about the solutions of (2.1). In the generic case, we introduce a one-parameter family of solutions of (2.1) by

$$\chi(x;a) = \begin{cases} m_l(0,x) + a\,m_r(0,x) & \text{if} \quad 0 \le a < \infty, \\ m_r(0,x) & \text{if} \quad a = \infty. \end{cases} \tag{2.2}$$

Let

$$\rho(x;a) = \frac{\chi'(x;a)}{\chi(x;a)}. \tag{2.3}$$

Then $\rho(x;a)$ satisfies the Riccati equation

$$V(x) = \rho'(x;a) + \rho(x;a)^2. \tag{2.4}$$

If $\psi(k,x)$ is a solution of (1.1), then

$$\phi(k,x;a) = \psi'(k,x) - \rho(x;a)\,\psi(k,x) \tag{2.5}$$

is a solution of (1.1) with the new potential

$$U(x;a) = \rho(x;a)^2 - \rho'(x;a). \tag{2.6}$$

In the exceptional case we define $\chi(x) = m_l(0,x)$ and $\rho(x) = m_l'(0,x)/m_l(0.x)$ for, since $m_l(0,x)$ and $m_r(0,x)$ are linearly dependent, the parameter a becomes superfluous. Applying the transformation (2.5) to the Jost solutions $f_l(k,x)$ and $f_r(k,x)$ associated with $V(x)$, we find for the Jost solutions $g_l(k,x;a)$ and $g_r(k,x;a)$ associated with $U(x;a)$ that

$$g_l(k,x;a) = \frac{1}{ik}\,[f_l'(k,x) - \rho(x;a)\,f_l(k,x)], \tag{2.7}$$

$$g_r(k,x;a) = -\frac{1}{ik}\,[f_r'(k.x) - \rho(x;a)\,f_r(k,x)], \tag{2.8}$$

and hence

$$n_l(k,x;a) = \frac{1}{ik}\,[m_l'(k,x) + ik\,m_l(k,x) - \rho(x;a)\,m_l(k,x)], \tag{2.9}$$

$$n_r(k,x;a) = -\frac{1}{ik}\,[m_r'(k,x) - ik\,m_r(k,x) - \rho(x;a)\,m_r(k,x)]. \tag{2.10}$$

We will refer to the transformations from ψ to ϕ given by (2.5) and from V to U given by (2.6) as Darboux transformations. It is straightforward to verify that the scattering matrix corresponding to $U(x;a)$ is $\mathbf{J}\,\mathbf{S}(k)\,\mathbf{J}$. The potentials U that arise as the Darboux transformation of potentials in L_1^1 can be completely characterized. For similar results, see [7] and [11]. Let $\mathcal{U}$ be the family of potentials U that satisfy the following three conditions

(i) U is of the form

$$U_{\epsilon_1,\epsilon_2}(x) = \epsilon_1 \frac{2}{x^2+1} H(x) + \epsilon_2 \frac{2}{x^2+1} H(-x) + W(x), \tag{2.11}$$

where $W \in L_1^1$ and $\epsilon_j = 0, 1$ $(j = 1, 2)$.

(ii) The Schrödinger equation with potential U has no bound states,

(iii) $k = 0$ is either a bound state or a half-bound state for the potential U.

We recall that $k = 0$ is a half-bound state when the Schrödinger equation has a zero-energy solution which is bounded but not in L^2. For given ϵ_1 and ϵ_2 we denote by $\mathcal{U}_{\epsilon_1,\epsilon_2}$ the class of potentials of the form $U_{\epsilon_1,\epsilon_2}$. Concerning condition (iii) we remark that $k = 0$ is a bound state if and only if $\epsilon_1 = \epsilon_2 = 1$ [1]. The connection between the class L_1^1 and the class $\mathcal{U}$ is made precise in the next theorem which is a compact version of Theorem 2.7 of [1]. Let $\mathcal{V}_g$ ($\mathcal{V}_e$) denote the class of generic (exceptional) potentials in L_1^1 without bound states. Since in the generic case the parameter a enters into the Darboux transformation, we also introduce the set $\mathcal{V}_g \times [0,\infty]$ whose elements we denote by $[V,a]$, $V \in \mathcal{V}_g$, $0 \le a \le \infty$. Consider the map $\mathcal{D} : (\mathcal{V}_g \times [0,\infty]) \cup \mathcal{V}_e \to \mathcal{U}$ defined by $\mathcal{D}([V,a]) = U(\cdot\,;a)$ if $V \in \mathcal{V}_g$ and $\mathcal{D}(V) = U(\cdot)$ if $V \in \mathcal{V}_e$. Then we have

Theorem 2.1 Suppose V has no bound states. $\mathcal{D}$ is one-to-one and onto. Specifically, $\mathcal{D}(\mathcal{V}_g \times \{0\}) = \mathcal{U}_{0,1}$, $\mathcal{D}(\mathcal{V}_g \times \{\infty\}) = \mathcal{U}_{1,0}$, $\mathcal{D}(\mathcal{V}_g \times (0,\infty)) = \mathcal{U}_{1,1}$ and $\mathcal{D}(\mathcal{V}_e) = \mathcal{U}_{0,0}$.

Associated with a potential $U(x;a)$ there is a solution $n(k,x;a)$ of (1.12) and hence a solution $\mathbf{M}(k,x;a)$ of (1.16). This establishes, in the absence of bound states, the existence of solutions for the Riemann-Hilbert problem (1.16). In the generic case we obtain a one-parameter family of solutions, whereas in the exceptional case we obtain just one solution. We emphasize that at this point we do not yet know whether there are any solutions of (1.16) other than the ones just described. We will refer to the solutions of (1.16) obtained so far as 'standard solutions'. We remark that in the generic case the matrix $\mathbf{M}(k,x;a)$ has a $1/k$-singularity at $k = 0$. Explicitly, we can show that as $k \to 0$ in $\overline{\mathbf{C}^+}$

$$\lim_{k\to 0} k\,\mathbf{M}(k,x;a) = \begin{cases} \dfrac{ic_r}{2\chi(x;a)} \begin{bmatrix} a & -a \\ -1 & 1 \end{bmatrix}, & 0 \le a < \infty, \\[2ex] \dfrac{ic_r}{2\chi(x;\infty)} \begin{bmatrix} 1 & -1 \\ 0 & 0 \end{bmatrix}, & a = \infty. \end{cases} \tag{2.12}$$

Thus we see that the parameter a can be recovered from the small-k behavior of $\mathbf{M}(k,x;a)$. If $0 < a < \infty$, then $\eta(x;a) = 1/\chi(x;a)$ is the zero-energy L^2-solution. Moreover, a is related to the norming constant corresponding to $\eta(x;a)$ by

$$\frac{1}{\int_{-\infty}^{\infty} \eta(x;a)^2\,dx} = a\,c_r. \tag{2.13}$$

In the exceptional case $\mathbf{M}(k,x)$ is continuous at $k=0$. We also mention that in both the exceptional and generic case $\det \mathbf{M}(k,x) = 1/T(k)$, which implies that $\mathbf{M}(k,x)^{-1}$ exists in $\overline{\mathbf{C}^+} \setminus \{0\}$ and is continuous at $k=0$. For the above details, see [1].

Next we consider the case when there are bound states $-\beta_{\mathcal{N}}^2 < \ldots < -\beta_1^2 < 0$. Then $T(k)$ has simple poles at $k = i\beta_j$ $(\beta_j > 0)$, $j = 1, \ldots, \mathcal{N}$. Define

$$\check{T}(k) = \left(\prod_{j=1}^{\mathcal{N}} \frac{k - i\beta_j}{k + i\beta_j} \right) T(k), \tag{2.14}$$

$$\check{R}(k) = (-1)^{\mathcal{N}} \left(\prod_{j=1}^{\mathcal{N}} \frac{k - i\beta_j}{k + i\beta_j} \right) R(k). \tag{2.15}$$

$$\check{L}(k) = (-1)^{\mathcal{N}} \left(\prod_{j=1}^{\mathcal{N}} \frac{k - i\beta_j}{k + i\beta_j} \right) L(k). \tag{2.16}$$

The potential corresponding to $\check{T}(k)$ will be denoted by $\check{V}(x)$ and its Jost solutions by $\check{f}_l(k,x)$ and $\check{f}_r(k,x)$. The potential $\check{V}(x)$ has no bound states since $\check{T}(k)$ has no poles in the upper half-plane. It is well known [10] that $V(x)$ and the associated Jost solutions can be obtained from $\check{V}(x)$ and its Jost solutions inductively by adding bound states one by one. A similar construction can be carried out for the potential $U(x;a)$. We can start from $\check{U}(x;a)$, a Darboux transform of $\check{V}(x)$, and add bound states one by one following the method used for $V(x)$. Since $U(x;a)$ is not in L_1^1, the various step need to be justified. This is done in [1]. It turns out that the values of ϵ_1 and ϵ_2 in (2.11) do not change if bound states are added. Also, $k=0$ is a bound state (half-bound state) for $U(x;a)$ if and only if $k=0$ is a bound state (half-bound state) for $\check{U}(x;a)$. The $\mathcal{N}$ norming constants associated with $V(x)$ and the $\mathcal{N}$ norming constants associated with $U(x;a)$ form a set of $2\mathcal{N}$ parameters. Together with the parameter a we obtain, in the generic case, a $2\mathcal{N}+1$ - parameter family of solutions of the Riemann-Hilbert problem (1.61). In the exceptional case, we obtain a $2\mathcal{N}$- parameter family of solutions. As in the case without bound states, we will refer to the solutions of (1.16) obtained by the above construction as 'standard solutions'.

3. THE UNIQUENESS PROBLEM

In this section we start from a solution of (1.16) and investigate whether it can be associated with potentials $V(x)$ and $U(x)$ and, if so, with what kind of potentials. Our assumptions on $\mathbf{M}(k,x)$ are

(i) $\mathbf{M}(k,x)$ is analytic in $\mathbf{C}^+$ and continuous in $\overline{\mathbf{C}^+} \setminus \{0\}$ for each x.

(ii) $\mathbf{M}(k,x) \to \mathbf{I}$ as $k \to +\infty$ in $\overline{\mathbf{C}^+}$ for every x, and $\mathbf{M}(-k,x) = \overline{\mathbf{M}(k,x)}$ for $k \in \mathbb{R}$.

(iii) $\mathbf{M}(k,x) = O(k^{-2})$ as $k \to 0$ in $\overline{\mathbf{C}^+}$ for every x.

Theorem 3.1 Suppose that $\mathbf{M}(k,x)$ is a solution of (1.16) which satisfies (i), (ii) and (iii) and where the scattering matrix comes from an L_1^1-potential without bound states. Then the only solutions of (1.16) that can be associated with a Schrödinger equation are of the standard type.

In [1], Theorem 3.1, the stronger assumption $\mathbf{M}(k,x) = O(k^{-1})$ was made in place of (iii). In the proof of Theorem 3.1 we will concentrate on how to deal with condition (iii).

The other aspects of the proof are similar as in the proof of [1]. Referring back to (1.9), we also remark that it follows as in [1] that if only one component of $m(k,x)$ ($n(k,x)$) is associated with a potential $V(x)$ ($U(x)$) then the other component is automatically associated with the same potential.

Proof: In this proof we denote the potential whose scattering matrix enters into (1.16) by $V_0(x)$ and we use a subscript zero to denote other quantities related to $V_0(x)$. Assume that $V_0(x)$ is generic. Everything goes through in essentially the same way if $V(x)$ is exceptional. We will comment on the exceptional case only where necessary. Now fix $b \in (0,\infty)$. Let $\mathbf{M}_0(k,x;b)$ denote the corresponding standard solution of (1.16), where b is the parameter in $U_0(x;b)$. The solution $\mathbf{M}_0(k,x;b)$ provides us with a matrix factorization of $\mathbf{G}(k,x)$, namely

$$\mathbf{G}(k,x) = \mathbf{M}_0(-k,x;b)\left[\mathbf{q}\mathbf{M}_0(k,x;b)^{-1}\mathbf{q}\right], \tag{3.1}$$

where the first factor has an analytic continuation to $\overline{\mathbf{C}^-}\setminus\{0\}$ and the second factor (in brackets) has an analytic extension to $\overline{\mathbf{C}^+}\setminus\{0\}$. Recall that the factor $\mathbf{M}_0(-k,x;b)$ has an $O(1/k)$-singularity at $k=0$ while the second factor is continuous at $k=0$. Let $\mathbf{M}(k,x)$ be an arbitrary solution of (1.16) which satisfies (i),(ii) and (iii). Then we can write (1.16) as

$$\mathbf{M}_0(-k,x;b)^{-1}\mathbf{M}(-k,x) = \mathbf{q}\,\mathbf{M}_0(k,x;b)^{-1}\mathbf{M}(k,x)\,\mathbf{q}. \tag{3.2}$$

Both sides of (3.2) are $O(k^{-2})$ as $k\to 0$. Hence, by a variant of Liouville's theorem both sides must be equal to

$$-\frac{\mathbf{A}_2(x)}{k^2}+\frac{i\mathbf{A}_1(x)}{k}+\mathbf{I} \tag{3.3}$$

for some matrix functions $\mathbf{A}_1(x)$ and $\mathbf{A}_2(x)$. By (3.2) and assumption (ii), we have that $\mathbf{q}\mathbf{A}_1(x)\mathbf{q} = -\mathbf{A}_1(x)$, $\overline{\mathbf{A}_1(x)} = \mathbf{A}_1(x)$, $\mathbf{q}\mathbf{A}_2(x)\mathbf{q} = \mathbf{A}_2(x)$ and $\overline{\mathbf{A}_2(x)} = \mathbf{A}_2(x)$. Hence $\mathbf{A}_1(x)$ and $\mathbf{A}_2(x)$ are of the form

$$\mathbf{A}_1(x) = \begin{bmatrix} a_1(x) & b_1(x) \\ -b_1(x) & -a_1(x)\end{bmatrix}, \tag{3.4}$$

$$\mathbf{A}_2(x) = \begin{bmatrix} a_2(x) & b_2(x) \\ b_2(x) & a_2(x)\end{bmatrix}, \tag{3.5}$$

where $a_j(x)$ and $b_j(x)$ ($j=1,2$) are real functions. From (3.2), (3.3), (1.13), (1.14) and (1.15) it follows that

$$m(k,x) = m_0(k,x)\left[1-\frac{\gamma_2(x)}{k^2}\right]+\frac{i}{k}\gamma_1(x)\,\mathbf{J}\,n_0(k,x;b), \tag{3.6}$$

$$n(k,x) = n_0(k,x;b)\left[1-\frac{\theta_2(x)}{k^2}\right]+\frac{i}{k}\theta_1(x)\,\mathbf{J}\,m_0(k,x), \tag{3.7}$$

where $\gamma_1(x) = -a_1(x)-b_1(x)$, $\gamma_2(x) = a_2(x)+b_2(x)$, $\theta_1(x) = b_1(x)-a_1(x)$ and $\theta_2(x) = a_2(x)-b_2(x)$. We see from (3.7) that in order to satisfy condition (iii) in the generic case, we must have that $\theta_2(x)=0$ [1]. We ignore this restriction for the moment. Now we insert (3.6) and (3.7) into (1.6) and (1.12) respectively, where the potentials $V(x)$ and $U(x)$ are viewed as unknowns. By using (2.9) and (2.10) and looking at the magnitudes of the terms as $k\to\infty$ ($k\in\mathbb{R}$), we obtain the following equations ($\rho_0=\rho_0(x;b)$)

$$V(x) = V_0(x) - 2\gamma_1'(x), \tag{3.8}$$

$$U(x) = U_0(x;b) - 2\theta_1'(x), \tag{3.9}$$

$$-2\gamma_2' + \gamma_1'' + 2\gamma_1'\gamma_1 - 2(\gamma_1\rho_0)' = 0. \tag{3.10}$$

$$\gamma_2'' + 2\gamma_1'\gamma_2 - 2\gamma_2'\rho_0 = 0, \tag{3.11}$$

$$-2\theta_2' + \theta_1'' + 2\theta_1'\theta_1 + 2(\theta_1\rho_0)' = 0. \tag{3.12}$$

$$\theta_2'' + 2\theta_1'\theta_2 - 2\theta_2'\rho_0 = 0. \tag{3.13}$$

We emphasize that the equations (3.8)–(3.13) are both necessary and sufficient for $m(k,x)$ and $n(k,x)$ to be solutions of a Schrödinger equation. Since $m(k,x)$ and $n(k,x)$ are related to the Jost solutions of $V(x)$ and $U(x)$ as described in Section 1, we must have that $m_l(k,x)$ and $n_l(k,x) \to 1$ as $x \to +\infty$, and $m_r(k,x)$ and $n_r(k,x) \to 1$ as $x \to -\infty$. Consequently, we require that

$$\gamma_j(x), \theta_j(x) \to 0 \qquad \text{as} \qquad x \to \pm\infty, \qquad (j = 1,2). \tag{3.14}$$

By integrating (3.10) and (3.12) once, we see that (3.14) implies $\gamma_j'(x) \to 0$ and $\theta_j'(x) \to 0$ as $x \to \pm\infty, (j = 1,2)$. We first consider (3.10) and (3.11). The solutions $\gamma_1(x)$ and $\gamma_2(x)$ can be expressed in the form

$$\gamma_1(x) = \frac{1}{u(x)}, \tag{3.15}$$

$$\gamma_2(x) = \frac{1}{u(x)\,w(x)}, \tag{3.16}$$

where $w(x)$ and $u(x)$ obey

$$w' - 2\rho_0(x;b)\,w = 1, \tag{3.17}$$

$$u' + 2\tilde{\rho}_0(x)\,u = 1, \tag{3.18}$$

and where $\tilde{\rho}_0(x)$ is related to $\rho_0(x;b)$ by

$$\tilde{\rho}_0(x) = \rho_0(x;b) + \frac{1}{w(x)}. \tag{3.19}$$

Solving (3.17) yields

$$w(x) = \chi_0(x;b)^2 \left[\int_0^x \chi_0(y;b)^{-2}\,dy + c_1\right]. \tag{3.20}$$

where c_1 is an arbitrary constant. To solve (3.18), let

$$\xi_0(x) = \chi_0(x;b) \left[\int_0^x \chi_0(y;b)^{-2}\,dy + c_1\right]. \tag{3.21}$$

and note that $\xi_0(x)$ is a solution of (2.1) with potential $V_0(x)$ which is linearly independent of $\chi_0(x;b)$. Furthermore,

$$\tilde{\rho}_0(x) = \frac{\xi_0'(x)}{\xi_0(x)}. \tag{3.22}$$

Then the solution of (3.18) is given by

$$u(x) = \xi_0(x)^{-2} \left[\int_0^x \xi_0(y)^2 \, dy + c_2 \right], \tag{3.23}$$

where c_2 is arbitrary. Since in the generic case $\chi_0(y;b)$ grows linearly as $y \to \pm\infty$, we see that the integral inside the brackets in (3.21) has finite limits as $x \to \pm\infty$. Let

$$d_\pm = \pm \int_0^{\pm\infty} \chi_0(y;b)^{-2} \, dy. \tag{3.24}$$

If $c_1 \geq d_-$ or $c_1 \leq -d_+$, then $w(x)$ and $\xi_0(x)$ have no zero but the term in brackets in (3.23) is zero for some unique value $x = x_0$. To see this, note that $\xi_0(x)$ grows linearly as $x \to +\infty$ if $c_1 \geq d_-$ or $c_1 < -d_+$ and approaches a nonzero limit if $c_1 = -d_+$. Similarly $\xi_0(x)$ grows linearly as $x \to -\infty$ if $c_1 \leq -d_+$ or $c_1 > d_-$ and approaches a nonzero limit if $c_1 = d_-$. Hence $u(x_0) = 0$ and so $\gamma_1(x)$ becomes singular as $x \to x_0$. This is unacceptable. If $c_1 \in (-d_+, d_-)$, then there is a unique x_1 such that the bracket in (3.20) vanishes. Hence $w(x_1) = \xi_0(x_1) = 0$. So either $x_0 \neq x_1$ in which case $u(x_0) = 0$ immediately, or $x_0 = x_1$ in which case $u(x) = c(x - x_0)^2 + o(x - x_0)^2$ $(c \neq 0)$ as $x \to x_0$ by expanding $u(x)$ near $x = x_0$. Thus in all cases $\gamma_1(x)$ is singular at x_0. One can also see that $\gamma_2(x)$ is singular at x_0. The cases where one of $\gamma_1(x)$ or $\gamma_2(x)$ is identically zero and the other is nonzero (formally $c_1 = \infty$ or $c_2 = \infty$) can also easily be ruled out. Hence we have that $\gamma_1(x) = \gamma_2(x) = 0$ for all x is the only acceptable solution. Hence by (3.6), $V(x) = V_0(x)$. In the exceptional case $\chi_0(x;b) = \chi_0(x)$ is independent of b and has nonzero limits as $x \to \pm\infty$. $\xi_0(x)$ still grows linearly at both $\pm\infty$. Hence the above reasoning leads again to the conclusion that $\gamma_1(x) = \gamma_2(x) = 0$. The system (3.12), (3.13) for $\theta_1(x)$ and $\theta_2(x)$ can be discussed in a similar way. In the generic case, we have already mentioned below (3.7) that $\theta_2(x) = 0$ at the outset. The remaining equation for $\theta_1(x)$ has been solved in [1]. One finds that $\theta_1(x) = \rho_0(x;a) - \rho_0(x;b)$ for some $a \in [0,\infty]$. This implies by (3.9), (2.4) and (2.6) that $U(x) = U_0(x;b) - 2\rho_0'(x;a) + 2\rho_0'(x;b) = V(x) - 2\rho_0'(x;a) = U_0(x;a)$. In the exceptional case one finds $\theta_1(x) = \theta_2(x) = 0$. Theorem 3.1 is proved. ∎

In regard to assumption (iii), we conjecture that Theorem 3.1 still holds if we assume that $\mathbf{M}(k,x) = O(k^{-n})$ for some $n \geq 3$ as $k \to 0$, or if we do not make any such assumption at all, in which case (3.3) would have to be replaced by a full Laurent expansion.

Now we turn to the case when V has $\mathcal{N}$ bound states $-\beta_\mathcal{N}^2 < \ldots < -\beta_1^2 < 0$. In addition to (1.16) we consider the Riemann-Hilbert problem

$$\check{\mathbf{M}}(-k,x) = \check{\mathbf{G}}(k,x)\, \mathbf{q}\, \check{\mathbf{M}}(k,x)\, \mathbf{q}, \tag{3.25}$$

where

$$\check{\mathbf{G}}(k,x) = \begin{bmatrix} \check{T}(k) & -\check{R}(k)\, e^{2ikx} \\ -\check{L}(k)\, e^{-2ikx} & \check{T}(k) \end{bmatrix}, \tag{3.26}$$

with $\check{T}(k)$, $\check{R}(k)$ and $\check{L}(k)$ given by are given by (2.14)–(2.16). Put

$$\check{\mathbf{M}}(k,x) = \frac{1}{2} \begin{bmatrix} \check{m}_l(k,x) + \check{n}_l(k,x) & \check{m}_l(k,x) - \check{n}_l(k,x) \\ \check{m}_r(k,x) - \check{n}_r(k,x) & \check{m}_r(k,x) + \check{n}_r(k,x) \end{bmatrix}, \tag{3.27}$$

such that $\check{m}(k,x) = \check{\mathbf{M}}(k,x)\,\hat{1}$ and $\check{n}(k,x) = \mathbf{J}\,\check{\mathbf{M}}(k,x)\,\hat{e}$. Then $\check{m}(k,x)$ and $\check{n}(k,x)$ are solutions of (1.6) and (1.12). Also $\check{\chi}(x)$ and $\check{\rho}(x)$ are the corresponding quantities given by

(2.2) and (2.3) with potentials $\check{V}(x)$ and $\check{U}(x)$, respectively. Also $\det \check{\mathbf{M}}(k,x) = 1/\check{T}(k)$ and hence $\check{\mathbf{M}}(k,x)$ is invertible for $k \in \overline{\mathbf{C}^+} \setminus \{0\}$. Similarly as in the proof of Theorem 2.2 the solution of the Riemann-Hilbert problem (3.25) provides us with a factorization of $\check{\mathbf{G}}(k,x)$, namely

$$\check{\mathbf{G}}(k,x) = \check{\mathbf{M}}(-k,x)\ \left[\mathbf{q}\,\check{\mathbf{M}}(k,x)^{-1}\,\mathbf{q}\right], \qquad k \in \mathbb{R}. \tag{3.28}$$

In the generic case, we can assume without loss that the parameter a which specifies $\check{U}(x)$ satisfies $0 < a < \infty$. We will suppress a from our notation. Moreover, note that $\mathbf{G}(k,x)$ and $\check{\mathbf{G}}(k,x)$ are related by

$$\mathbf{G}(k,x) = \left(\prod_{j=1}^{\mathcal{N}} \frac{k+i\beta_j}{k-i\beta_j}\right) \mathbf{J}^{\mathcal{N}}\,\check{\mathbf{G}}(k,x)\,\mathbf{J}^{\mathcal{N}}. \tag{3.29}$$

It is convenient to define

$$\mathbf{N}(k,x) = \mathbf{J}^{\mathcal{N}}\,\check{\mathbf{M}}(k,x)\,\mathbf{J}^{\mathcal{N}}. \tag{3.30}$$

Inserting (3.29) in (1.16) and using (3.28) and (3.30), we obtain for $k \in \mathbb{R}$

$$\left(\prod_{j=1}^{\mathcal{N}}(k-i\beta_j)\right)\mathbf{N}(-k,x)^{-1}\,\mathbf{M}(-k,x) = \left(\prod_{j=1}^{\mathcal{N}}(k+i\beta_j)\right)\mathbf{q}\,\mathbf{N}(k,x)^{-1}\,\mathbf{M}(k,x)\,\mathbf{q}. \tag{3.31}$$

Here we impose the condition that

$$\mathbf{M}(k,x) = O(1/k) \qquad \text{as} \qquad k \to 0 \qquad \text{in} \qquad \overline{\mathbf{C}^+}, \tag{3.32}$$

which is the same condition as in [1]. We have not yet studied weaker conditions like $\mathbf{M}(k,x) = O(k^{-2})$. in connection with bound states. Since $\check{\mathbf{M}}(k,x)^{-1}$ and hence $\mathbf{N}(k,x)^{-1}$ are continuous at $k=0$, an appeal to Liouville's theorem yields that both sides of (3.31) must be equal to a matrix function in k of the form

$$k^{\mathcal{N}}\left[\mathbf{I} + \sum_{j=1}^{\mathcal{N}+1}(i/k)^j\mathbf{A}_j(x)\right]. \tag{3.33}$$

Thus, by (3.31),

$$\mathbf{M}(k,x) = \frac{k^{\mathcal{N}}}{\prod_{j=1}^{\mathcal{N}}(k+i\beta_j)}\mathbf{N}(k,x)\left[\mathbf{I} + \sum_{j=1}^{\mathcal{N}+1}(i/k)^j\,\mathbf{q}\,\mathbf{A}_j(x)\,\mathbf{q}\right]. \tag{3.34}$$

Since $\mathbf{N}(-k,x) = \overline{\mathbf{N}(k,x)}$ and $\mathbf{M}(-k,x) = \overline{\mathbf{M}(k,x)}$ for $k \in \mathbb{R}$, we conclude that the matrices $\mathbf{A}_j(x)$ are real and that $\mathbf{A}_j(x) = (-1)^j\mathbf{q}\,A_j(x)\,\mathbf{q}$. In other words, $\mathbf{A}_j(x)$ is of the form

$$\mathbf{A}_j(x) = \begin{bmatrix} a_j(x) & b_j(x) \\ b_j(x) & a_j(x) \end{bmatrix}, \qquad j \text{ even}, \tag{3.35}$$

and

$$\mathbf{A}_j(x) = \begin{bmatrix} a_j(x) & b_j(x) \\ -b_j(x) & -a_j(x) \end{bmatrix}, \qquad j \text{ odd}, \tag{3.36}$$

where $a_j(x)$ and $b_j(x)$ are real functions. Let $\gamma_j(x) = (-1)^j [a_j(x) + b_j(x)]$ and $\theta_j(x) = (-1)^j [a_j(x) - b_j(x)]$. Then by using (3.34), (2.14) and (2.15) we get

$$m(k,x) = \mathbf{M}(k,x)\hat{1} = \frac{k^{\mathcal{N}}}{\prod_{j=1}^{\mathcal{N}}(k+i\beta_j)}\mathbf{N}(k,x)\left[\hat{1} + \sum_{j=1}^{\mathcal{N}+1}(i/k)^j \gamma_j(x)\mathbf{J}^j\hat{1}\right], \tag{3.37}$$

$$n(k,x) = \mathbf{J}\mathbf{M}(k,x)\hat{e} = \frac{k^{\mathcal{N}}}{\prod_{j=1}^{\mathcal{N}}(k+i\beta_j)}\mathbf{J}\,\mathbf{N}(k,x)\left[\hat{e} + \sum_{j=1}^{\mathcal{N}+1}(i/k)^j \theta_j(x)\mathbf{J}^j\hat{e}\right]. \tag{3.38}$$

Looking at (3.38) as $k \to 0$, we see that in the generic case in order to satisfy (3.32), we must have that $\gamma_{\mathcal{N}+1}(x) = 0$. Inserting (3.37) and (3.38) in (1.6) and (1.12) respectively, and letting $k \to \infty$ leads to the following equations (see [1] for more details)

$$V(x) = \begin{cases} \check{V}(x) - 2\gamma_1'(x), & \mathcal{N} \text{ even}, \\ \check{U}(x) - 2\gamma_1'(x), & \mathcal{N} \text{ odd}, \end{cases} \tag{3.39}$$

$$U(x) = \begin{cases} \check{U}(x) - 2\theta_1'(x), & \mathcal{N} \text{ even}, \\ \check{V}(x) - 2\theta_1'(x), & \mathcal{N} \text{ odd}. \end{cases} \tag{3.40}$$

In the equations (3.41)–(3.44) the upper (lower) sign refers to $\mathcal{N}$ even (odd).

$$-2\gamma_{2j}' + \gamma_{2j-1}'' + 2\gamma_1'\gamma_{2j-1} \mp 2(\gamma_{2j-1}\check{\rho})' = 0, \tag{3.41}$$

$$-2\gamma_{2j+1}' + \gamma_{2j}'' + 2\gamma_1'\gamma_{2j} \pm 2\gamma_{2j}'\check{\rho} = 0, \tag{3.42}$$

$$-2\theta_{2j}' + \theta_{2j-1}'' + 2\theta_1'\theta_{2j-1} \pm 2(\theta_{2j-1}\check{\rho})' = 0, \tag{3.43}$$

$$-2\theta_{2j+1}' + \theta_{2j}'' + 2\theta_1'\theta_{2j} \mp 2\theta_{2j}'\check{\rho} = 0. \tag{3.44}$$

In (3.41)–(3.44) we can let j range from 1 to $+\infty$ if we assume that $\gamma_j(x) = 0$ for $j > \mathcal{N}+1$ ($j > \mathcal{N}$ the generic case) and $\theta_j(x) = 0$ for $j > \mathcal{N}+1$. Since $m_l(k,x)$, $n_l(k,x) \to 1$ as $x \to +\infty$ and $m_r(k,x)$, $n_r(k,x) \to 1$ as $x \to -\infty$, it follows that $\gamma_j(x)$ and $\theta_j(x)$ must satisfy the following boundary conditions

$$\theta_j(+\infty) = \gamma_j(+\infty) = \sum_{i_1<\ldots<i_j} \beta_{i_1}\cdots\beta_{i_j}, \tag{3.45}$$

$$\theta_j(-\infty) = \gamma_j(-\infty) = (-1)^j \sum_{i_1<\ldots<i_j} \beta_{i_1}\cdots\beta_{i_j}, \tag{3.46}$$

for $j = 1, \ldots, \mathcal{N}$, and

$$\theta_{\mathcal{N}+1}(\pm\infty) = 0. \tag{3.47}$$

The further details of the analysis are to lengthy to be given here. We refer again to [1]. We mention, however, that the following identity holds

$$\theta_{\mathcal{N}}'\,\theta_{\mathcal{N}+1} - \theta_{\mathcal{N}}\,\theta_{\mathcal{N}+1}' - \theta_{\mathcal{N}+1}^2 - 2\,\check{\rho}\,\theta_{\mathcal{N}}\,\theta_{\mathcal{N}+1} = 0. \tag{3.48}$$

This allows us to introduce

$$w(x) = \frac{\theta_{\mathcal{N}}(x)}{\theta_{\mathcal{N}+1}(x)}, \tag{3.49}$$

where $w(x)$ satisfies

$$w' - 2\,\check{\rho} w = 1, \tag{3.50}$$

and hence

$$w(x) = \check{\chi}^2(x)\left[\int_{-\infty}^{x} \check{\chi}(y)^{-2}\,dy + c\right]. \tag{3.51}$$

Of course, (3.48) is also satisfied if $\theta_{\mathcal{N}+1}(x) = 0$ indentically. The analysis is now divided into two parts. First one considers the solutions of (3.41)–(3.44) under the additional assumption that $\theta_{\mathcal{N}+1}(x) = 0$. Then one shows that one can reduce the case with $\theta_{\mathcal{N}+1}(x) \neq 0$ to that with $\theta_{\mathcal{N}+1}(x) = 0$. We only state the main results. The proofs are given in [1].

Proposition 3.2 $\gamma_{\mathcal{N}+1}(x) = \theta_{\mathcal{N}+1}(x) = 0$ if and only if $\mathbf{N}(k,x)^{-1}\,\mathbf{M}(k,x)$ is continuous at $k = 0$.

Proposition 3.3 For the standard solutions we have that $\theta_{\mathcal{N}+1}(x) = 0$.

In the following theorem it is assumed that in addition to the parameter a the norming constants associated with the eigenvalues $-\beta_{\mathcal{N}}^2 < \ldots < -\beta_1^2 < 0$ for both $V(x)$ and $U(x;a)$ have been specified. In the generic case we also know that $\gamma_{\mathcal{N}+1}(x) = 0$ (see below (3.38)).

Theorem 3.4 The equations (3.43)–(3.44) with the boundary conditions (3.45)–(3.46) have exactly one solution which also satisfies $\theta_{\mathcal{N}+1}(x) = 0$. Similarly, the equations (3.41)–(3.42) together with (3.45)–(3.46) have exactly one solution which also satisfies $\gamma_{\mathcal{N}+1}(x) = 0$. The corresponding solutions of (1.16) are of the standard type. Furthermore, the solutions with $\theta_{\mathcal{N}+1}(x) \neq 0$ and for which $w(x) \neq 0$ are also of the standard type.

We remark that in the exceptional case for $U(x;a)$, the function $w(x)$ in (3.51) has a zero since $\check{\chi}(x)$ approaches finite nonzero limits as $x \to \pm\infty$. A similar result holds with respect to $V(x)$ and the system (3.41)–(3.42). In the exceptional case our analysis is not yet complete. We have so far only analyzed a special case and we briefly describe the results here. Let

$$T(k) = \frac{k + i\beta}{k - i\beta} \qquad \text{and} \qquad R(k) = L(k) = 0. \tag{3.52}$$

The corresponding potential is given in [1]. In this case $\check{V}(x) = \check{U}(x) = 0$ and hence $\check{\rho}(x) = 0$ and $\check{\chi}(x) = 1$, so we have the simplest possible case of the equations (3.43) and (3.44), namely

$$-2\theta_2' + \theta_1'' + 2\theta_1'\,\theta_1 = 0, \tag{3.53}$$

$$\theta_2'' + 2\,\theta_1'\theta_2 = 0. \tag{3.54}$$

Integrating the first equation, using the boundary conditons $\theta_1(\pm\infty) = \pm\beta$, $\theta_2(\pm\infty) = 0$, gives

$$-2\theta_2 + \theta_1' + \theta^2 = \beta^2. \tag{3.55}$$

From (3.51), we have that $w(x) = x + c$. Without loss of generality we set $c = 0$, since $c \neq 0$ can be dealt with by a shift of the origin. Hence

$$\theta_2(x) = \frac{\theta_1(x)}{x}. \tag{3.56}$$

Inserting (3.56) in the integrated form of (3.53) and making the substitution

$$u(x) = \theta_1(x) - \frac{1}{x}. \tag{3.57}$$

yields

$$u' + u^2 = \beta^2 + \frac{2}{x^2}. \tag{3.58}$$

So, if we set $u = \frac{\phi'}{\phi}$, then ϕ obeys

$$\phi'' = (\beta^2 + \frac{2}{x^2})\,\phi. \tag{3.59}$$

We are looking for solutions that are nonzero and finite for $x \neq 0$ and for which $u(\pm\infty) = \pm\beta$. Choosing the solutions to be positive we have

$$\phi(x) = \begin{cases} (1 + \frac{1}{\beta x})\,e^{-\beta x} + a\,(1 - \frac{1}{\beta x})\,e^{\beta x}, & x > 0, \\ (1 + \frac{1}{\beta x})\,e^{-\beta x} + b\,(1 - \frac{1}{\beta x})\,e^{\beta x}, & x < 0, \end{cases} \tag{3.60}$$

where $0 < a \leq 1$ and $b \geq 1$. One can see that the values $a = 1$ and $b = 1$ must be excluded because of the behavior of the resulting potential at $x = 0$. Hence we have that $0 < a < 1$ and $b > 1$. Now $\theta_1(x)$ and $\theta_2(x)$ are determined. Expanding them for small x one finds that $\theta_1(x)$, $\theta_1'(x)$ and $\theta_2(x)$ are continuous at $x = 0$, but that $\theta_2'(x)$ has a jump. Explicitly,

$$\theta_2'(0+) - \theta_2'(0-) = \frac{2\beta^3\,(b-a)}{(1-a)(b-1)}. \tag{3.61}$$

Since

$$n(k,x) = \frac{k}{k+i\beta}[\hat{1} + \frac{i}{k}\,\theta_1(x)\,\hat{e} - \frac{\theta_2(x)}{k^2}\,\hat{1}], \tag{3.62}$$

by using (3.61) and (1.12), we can see that the following potential would accomodate the discontinuity of $\theta_2'(x)$ at $x = 0$,

$$U(x) = -2\theta_1'(x) - \frac{2\,\beta^3\,(b-a)}{(1-a)\,(b-1)\,(k^2+\beta^2)}\,\delta(x). \tag{3.63}$$

To be sure, this is an absurd potential. We suspect that it is indicative of what happens in other cases when $w(x)$ has a zero.

Acknowledgements: The author is indebted to Y. A. Mishev for making the reference [11] available to him.

REFERENCES

1. T. Aktosun, M. Klaus and C. van der Mee, preprint (1992)

2. V. A. Marchenko, Sturm-Liouville Operators and Applications, Birkhäuser, Basel-Bonn-Stuttgart (1986).

3. M. Klaus, Inverse Problems, 4, 505 (1988).

4. R. G. Newton, J. Math. Phys., 21, 493 (1980).

5. T. Aktosun and R. G. Newton, Inverse Problems, 1, 291 (1985).

6. T. Aktosun, Inverse Problems, 3, 523 (1987).

7. A. Degasperis and P. C. Sabatier, Inverse Problems, 3, 73 (1987).

8. T. Aktosun, Thesis, Indiana University, Bloomington, unpublished (1986).

9. T.Aktosun, Phys. Rev. Lett., 58, 2159 (1987).

10. P. Deift and E. Trubowitz, Comm. Pure Appl. Math., 32, 121 (1979).

11. Y. P. Mishev, Inverse Problems, 7, 379 (1991).

Third International Colloquium on Differential Equations pp. 147-156 (1993)
D. Bainov and V. Covachev (Eds)

Optimal Control for Nonlinear Systems of Partial Differential Equations Related to Ecology

Anthony W. Leung
Department of Mathematical Sciences
University of Cincinnati
Cincinnati, OH, 45221-0025, U.S.A.

1 Introduction and Preliminary Results for Scalar Equations

This article summarizes some recent results concerning the optimal control of systems of nonlinear partial differential equations related to the harvesting of interacting populations. We will describe some steady state control and time-periodic control of prey-predator and competing interacting species. First, we consider a simple preliminary problem concerning the steady-state control of one species, whose growth is governed by the diffusive Volterra-Lotka equation with no-flux boundary condition:

$$(1.1)\qquad \Delta u + u[(a(x) - f(x)) - bu] = 0 \text{ in } \Omega, \quad \frac{\partial u}{\partial \nu} = 0 \text{ on } \partial\Omega.$$

The function $a(x)$ describes spatially dependent intrinsic growth rate, and b designates crowding effect which is assumed to be constant for simplicity. The function $f(x)$ denotes a distribution of control harvesting effort on the biological species. The optimal control criteria is to maximize the difference between economic revenue and cost. This is expressed by the payoff functional

$$(1.2)\qquad J(f) = \int_\Omega \{Kuf - Mf^2\}dx,$$

where K and M are constants describing the price of the species and the cost of the control. Here $\int_\Omega uf dx$ is a measure of the total harvest, and u itself depends on f through (1.1). Such a model is certainly only a prototype, and various other variations can be made.

To fixed ideas, we assume Ω is a bounded domain in R^n with $\partial\Omega \in C^2$; Δ and $\frac{\partial}{\partial \nu}$ respectively denote the Laplacian and outward normal derivative. K, M and b are positive constants. We also assume that

$$a(x) \geq 0, \;\; f(x) \geq 0 \text{ a.e. in } \Omega, \text{ and}$$

$$a \in L^\infty(\Omega), \quad f \in L^\infty(\Omega).$$

For convenience, we denote

$$L^\infty_+(\Omega) = \{f | f \in L^\infty(\Omega), \quad f \geq 0 \text{ a.e. in } \Omega\};$$

and for $\delta > 0$

(1.3) $$C_\delta = \{f | 0 \leq f \leq \delta \text{ a.e. in } \Omega\}.$$

We define an optimal control (if it exists) to be an $f^* \in C_\delta$ such that

(1.4) $$J(f^*) = \sup_{f \in C_\delta} J(f).$$

Before investigating the optimal control, we first note that for each fixed $f \in C_\delta$, problem (1.1) has a unique positive solution under appropriate assumptions.

Theorem 1.1 *Suppose that $a(x)$ and δ satisfy hypothesis*

[H1] $\quad 0 < \delta < \inf_\Omega a(x).$

Then for each $f \in C_\delta$, the problem (1.1) has a unique strictly positive solution $u \equiv u(f) \in W^{2,p}(\Omega)$, for any $p \in [1, \infty)$. Furthermore, the estimate

$$\|u(f)\|_{2,p} \leq const$$

is valid, uniformly for all $f \in C_\delta$.

This theorem is proved by the method of upper and lower solutions, using the fact that large positive constant and the constant

(1.5) $$\epsilon = \frac{1}{b}[\inf_\Omega a(x) - \delta] > 0$$

are respectively upper and lower solutions for (1.1). See [5] for details.

Since there are upper and lower bounds for all solutions of (1.1) uniformly for all $f \in C_\delta$, we can find a maximizing sequence $f_n \in C_\delta$ and prove in the usual manner that there exist a subsequence which converges weakly in $L^2(\Omega)$ to an optimal control $f^* \in C_\delta$.

Theorem 1.2 *Let δ and $a(x)$ satisfy hypothesis [H1], then an optimal control does exist.*

In order to characterize the optimal control, we next find a slightly stronger condition than [H1] to obtain differentiability of $u(f)$ with respect to f as described in the following lemma.

Lemma 1.1 *Suppose δ and $a(x)$ satisfy*

[H2] $$0 < \delta \leq \tfrac{1}{3}\{2\inf_\Omega a(x) - \sup_\Omega a(x)\}.$$

Then, the mapping

$$C_\delta \ni f \mapsto u(f) \in W^{1,2}(\Omega)$$

is differentiable in the following sense:

$$\frac{u(f+\beta\bar{f}) - u(f)}{\beta} \to \xi \quad \text{weakly in } W^{1,2}(\Omega)$$

as $\beta \to 0$, for any $f \in C_\delta$ and $\bar{f} \in L^\infty(\Omega)$ such that $f + \beta\bar{f} \in C_\delta$. Further, ξ is the unique solution of

$$\Delta\xi + (a - f - 2bu(f))\xi = \bar{f}u(f) \text{ in } \Omega, \quad \frac{\partial \xi}{\partial \nu} = 0 \text{ on } \partial\Omega. \tag{1.6}$$

We now obtain a characterization of the optimal control as follows.

Theorem 1.3 *Suppose $a(x)$ and δ satisfy hypothesis [H2]; and the constant K, M have the property:*

$$M \geq [K \sup_\Omega a]/(2b\delta). \tag{1.7}$$

Then for any optimal control $f \in C_\delta$, there exists (u,p) with $b^{-1}[\inf_\Omega a - \delta] \leq u \leq b^{-1}\sup_\Omega a$, $0 \leq p \leq K$, such that

$$f = \frac{u}{2M}(K - p) \text{ in } \Omega, \tag{1.8}$$

and (u,p) is a solution of the optimality system

$$\begin{cases} \Delta u + au - (b + \frac{K-p}{2M})u^2 = 0 \\ \Delta p + (a - 2bu)p + \frac{(K-p)^2 u}{2M} = 0 \text{ in } \Omega \\ \frac{\partial u}{\partial \nu} = \frac{\partial p}{\partial \nu} = 0 \text{ on } \partial\Omega. \end{cases} \tag{1.9}$$

Proof. The existence of an optimal control in C_δ has been justified by Theorem 1.2. Let $f \in C_\delta$ be an optimal control. For $\bar{g} \in L^\infty_+(\Omega)$, $\epsilon > 0$, set

$$\bar{f} = \bar{f}_\epsilon = \begin{cases} \bar{g} & \text{if } f \le \delta - \epsilon\|\bar{g}\|_\infty \\ 0 & \text{elsewhere.} \end{cases}$$

Then, for $\beta > 0$ small enough, we have

$$J(f) \ge J(f + \beta\bar{f}). \tag{1.10}$$

Dividing by β, and letting $\beta \to 0$, we use Lemma 1.1 to obtain

$$\int_\Omega Kf\xi + Ku(f)\bar{f} - 2Mf\bar{f}dx \le 0. \tag{1.11}$$

Now, define p to be the solution of

$$\Delta p + [a - f - 2bu(f)]p = -Kf \text{ in } \Omega, \ \frac{\partial p}{\partial \nu} = 0 \text{ on } \partial\Omega. \tag{1.12}$$

Note that from the lower bound (1.5) for $u(f)$, we have $a - f - 2bu(f) \le \sup_\Omega a - 2\inf_\Omega a + 2\delta \le -3\delta + 2\delta = -\delta$. Hence, the problem (1.12) has a unique solution p, and moreover

$$0 \le p \le K. \tag{1.13}$$

Combining (1.11) and (1.12), we integrate by parts and use Lemma 1.1 to obtain

$$\int_\Omega \bar{f}_\epsilon[u(f)(K - p) - 2Mf]dx \le 0. \tag{1.14}$$

Letting $\epsilon \to 0$, (1.14) leads to

$$\int_{\Omega\cap\{f<\delta\}} \bar{g}[u(f)(K - p) - 2Mf]dx \le 0 \ \forall \bar{g} \in L^\infty_+(\Omega). \tag{1.15}$$

Consequently, we must have

$$f \ge \frac{u(f)}{2M}(K - p) \text{ in } \Omega \cap \{f < \delta\}. \tag{1.16}$$

On the other hand, for $-\bar{g} \in L^\infty_+(\Omega)$, $\epsilon > 0$, we set

$$\bar{f} = \bar{f}_\epsilon = \begin{cases} \bar{g} & \text{if } f \ge \epsilon\|\bar{g}\|_\infty \\ 0 & \text{elsewhere.} \end{cases}$$

We deduce in the same way as above that (1.14) holds for such $\bar{f}_\epsilon$. Passing to the limit as $\epsilon \to 0$, we obtain an inequality analogous to (1.15) and deduce that

$$f \le \frac{u(f)}{2M}(K-p) \text{ in } \Omega \cap \{f>0\}. \tag{1.17}$$

Combining (1.16) and (1.17), and using (1.7), we can finally obtain

$$f = \frac{u(f)}{2M}(K-p) \text{ in } \Omega. \tag{1.18}$$

From (1.1), (1.12), and (1.18) we easily derive (1.9).

2 Optimal Control of Steady-State Prey-Predator Diffusive Volterra-Lotka Systems

This section considers the optimal harvesting control of two interacting populations. The species concentrations satisfy a predator-prey Volterra-Lotka system under diffusion. They are in steady-state situation with no-flux boundary conditions as follows:

$$\begin{aligned} \Delta u + u[(a_1(x) - f_1(x)) - b_1 u - c_1 v] &= 0 \text{ in } \Omega \\ \Delta v + v[(a_2(x) - f_2(x)) + c_2 u - b_2 v] &= 0 \text{ in } \Omega \\ \frac{\partial u}{\partial \nu} = \frac{\partial v}{\partial \nu} &= 0 \text{ on } \partial\Omega. \end{aligned} \tag{2.1}$$

The functions $u(x), v(x)$ respectively describe prey, predator population concentrations with intrinsic growth rates $a_1(x), a_2(x)$. The functions $f_1(x)$, $f_2(x)$ respectively denote distribution of control harvesting effect on the biological species. The parameters b_i, c_i, $i = 1,2$ are positive constants designating the crowding and interaction among the species. The optimal control criteria is to maximize economic return expressed by the payoff functional

$$J(f_1, f_2) = \int_\Omega \{K_1 u f_1 + K_2 v f_2 - M_1 f_1^2 - M_2 f_2^2\} dx \tag{2.2}$$

as in (1.2). Here K_1, K_2 are constants describing the price of the prey and predator species, and M_1, M_2 are constants describing the costs of the controls f_1, f_2.

As in Section 1, we assume

$$a_i(x) \geq 0, \quad f_i(x) \geq 0 \text{ a.e. in } \Omega, \quad \text{and}$$
$$a_i \in L^\infty(\Omega), \quad f_i \in L^\infty(\Omega), \quad i = 1, 2.$$

For $\delta_i > 0$, $i = 1, 2$, we denote

$$C(\delta_1, \delta_2) = \{(f_1, f_2) | 0 \leq f_i \leq \delta_i, \quad \text{a.e. in } \Omega, \quad i = 1, 2\};$$

and we define an optimal control (if it exists) to be an $(f_1^*, f_2^*) \in C(\delta_1, \delta_2)$ such that

$$J(f_1^*, f_2^*) = \sup\{J(f_1, f_2) | (f_1, f_2) \in C(\delta_1, \delta_2)\}.$$

Before obtaining the optimal control, we first deduce the existence and uniqueness of positive solution fo (2.1) for each fixed $(f_1, f_2) \in C(\delta_1, \delta_2)$. For convenience, we denote

$$\bar{a}_i = \sup_\Omega a_i(x), \quad \tilde{a}_i = \inf_\Omega a_i(x) \quad \text{for } i = 1, 2.$$

Theorem 2.1 *Suppose that $a_i(x), b_i, c_i$ and δ_i satisfy the hypotheses*

[P1] $\quad \tilde{a}_1 - \frac{c_1}{b_2}[\bar{a}_2 + \frac{c_2 \bar{a}_1}{b_1}] > \delta_1 > 0$, *and*

[P2] $\quad \tilde{a}_2 > \delta_2 > 0.$

Then for each pair $(f_1, f_2) \in C(\delta_1, \delta_2)$, the problem (2.1) has a strictly positive solution $(u, v) = (u(f_1, f_2), v(f_1, f_2))$, i.e. $u, v > 0$ in $\bar{\Omega}$, and with each component in $W^{2,p}(\Omega)$ for any $p \in [1, \infty)$. Moreover, the estimate

$$\|u(f_1, f_2)\|_{2,p}, \|v(f_1, f_2\|_{2,p} \leq const$$

is valid uniformly for all $(f_1, f_2) \in C(\delta_1, \delta_2)$.

Corollary 2.1 *Assume all the hypotheses of Theorem 2.1. Furthermore, if the quantity $c_1 c_2 (b_1 b_2)^{-1}$ is sufficiently small, then the problem (2.1) has a unique solution $u, v \in W^{2,p}(\Omega)$, $p \geq 1$, with the property:*

$$\phi_1 \leq u(x) \leq \psi_1, \quad \phi_2 \leq v(x) \leq \psi_2 \quad in\ \bar{\Omega}.$$

Here ϕ_i, ψ_i, $i = 1, 2$ are positive constants given by

$$
\begin{aligned}
\psi_1 &\equiv \bar{a}_1/b_1, \quad \psi_2 \equiv \frac{1}{b_1}[\bar{a}_2 + c_2\bar{a}_1/b_1], \\
\phi_1 &\equiv \frac{1}{b_2}[\tilde{a}_1 - \frac{c_1}{b_2}(\bar{a}_2 + c_2\bar{a}_1/b_1) - \delta_1], \quad \textit{and} \\
\phi_2 &\equiv \frac{1}{b_2}[\tilde{a}_2 - \delta_2 + \frac{c_2}{b_1}\{\tilde{a}_1 - \frac{c_1}{b_2}(\bar{a}_2 + c_2\bar{a}_1/b_1) - \delta_1\}].
\end{aligned} \tag{2.3}
$$

Theorem 2.1 is proved by means of a fixed point theorem and Corollary 2.1 is proved by using a contraction argument. Explicit condition concerning how small $c_1c_2(b_1b_2)^{-1}$ is sufficient for Corollary 2.1 to hold is given in [4].

Under the assumptions of Corollary 2.1, one can readily show as in Theorem 1.2 that an optimal control $(f_1^*, f_2^*) \in C(\delta_1, \delta_2)$ as described above does exist. Under more stringent conditions, it can be shown that the optimal control can be rigorously characterized. The conditions are elaborate because the interaction relationship between the two species is not symmetric. For notational convenience, we denote

$$E_1 = 2\tilde{a}_1 - \bar{a}_1 - \frac{2c_1}{b_2}(\bar{a}_2 + \frac{c_2\bar{a}_1}{b_1}), \quad E_2 = 2\tilde{a}_2 - \bar{a}_2 - \frac{c_2\bar{a}_1}{b_1}$$

Theorem 2.2 *Assume that*

[P3] $\quad 0 < \delta_i < \frac{1}{4}E_i, \ i = 1, 2,$

[P4] $\quad \frac{c_1\bar{a}_1}{b_1} + \frac{c_2}{b_2}(\bar{a}_2 + \frac{c_2\bar{a}_1}{b_1}) < 2\min\{\delta_1, \delta_2, 1\},$

[P5] $\quad M_1 > \frac{K_1\bar{a}_1}{2b_1\delta_1}[1 + 4\bar{a}_1c_1c_2(\bar{a}_2 + c_2\bar{a}_1/b_1)(b_1b_2E_1E_2)^{-1}],$ *and*

[P6] $\quad M_2 > \frac{1}{2b_2\delta_2}(\bar{a}_2 + c_2\bar{a}_1/b_1)[K_2 + 2\bar{a}_1c_1K_1(b_1E_2)^{-1}].$

Further, let c_1, c_2 be sufficiently small; and $(u(f_1, f_2), v(f_1, f_2))$ is uniquely defined for all $(f_1, f_2) \in C(\delta_1, \delta_2)$ in the sense described in Corollary 2.1. Suppose $(f_1^*, f_2^*) \in C(\delta_1, \delta_2)$ is an optimal control. Then

$$f_1^* = \frac{(K_1 - p_1)u}{2M_1}, \quad f_2^* = \frac{(K_2 - p_2)v}{2M_2}, \tag{2.4}$$

where (u, v, p_1, p_2) is a solution of the optimality system:

$$
(2.5)\quad \begin{cases}
\Delta u + [(a_1 - \frac{(K_1 - p_1)u}{2M_1}) - b_1 u - c_1 v]u = 0 \\
\Delta v[(a_2 - \frac{(K_2 - p_2)v}{2M_2}) + c_2 u - b_2 v]v = 0 \\
\Delta p_1 + a_1 p_1 + \frac{(K_1 - p_1)^2 u}{2M_1} - (2b_1 u + c_1 v)p_1 + c_1 v p_2 = 0, \ \textit{in } \Omega \\
\Delta p_2 + a_2 p_2 + \frac{(K_2 - p_2)^2 v}{2M_2} - (2b_2 v - c_2 u)p_2 - c_1 u p_1 = 0. \\
\frac{\partial u}{\partial \nu} = \frac{\partial v}{\partial \nu} = \frac{\partial p_1}{\partial \nu} = \frac{\partial p_2}{\partial \nu} = 0 \ \textit{on } \partial\Omega,
\end{cases}
$$

with each component in $H^{2,2}(\Omega)$ satisfying

$$
(2.6)\quad \begin{array}{l} \phi_1 \le u \le \psi_1, \ \ \phi_2 \le v \le \psi_2 \\ 0 \le p_1 \le K_1, \ \ 0 \le p_2 \le K_2/2 \end{array} \quad \textit{in } \Omega.
$$

Note that assumption [P3] implies [P1] and [P2] are valid.

Under the conditions of Theorem 2.2, it is possible to deduce a constructive method of approximating or computing the positive solution (u, v, p_1, p_2) of the optimality system (2.5). First, let

$$
\begin{gathered}
u_o \equiv \phi_1, \ \ u_{-1} \equiv \psi_1; \ \ v_0 \equiv \phi_2, \ \ v_{-1} \equiv \psi_2 \\
p_{1,C} \equiv 0, \ \ p_{1,-1} \equiv K_1; \ \ p_{2,0} \equiv 0, \ \ p_{2,-1} \equiv K_2/2
\end{gathered}
$$

Define $u_k(x), v_k(x), p_{1,k}(x), p_{2,k}(x)$ in Ω, $k = 1, 2, \ldots$ as solutions of scalar problems as follows:

$$
(2.7)\quad \begin{cases}
\Delta u_k - R u_k = -u_{k-2}[a_1 - \frac{(K_1 - p_{1,k-2})u_{k-2}}{2M_1} - b_1 u_{k-2} \\
\qquad -c_1 v_{k-1}] - R u_{k-2} \\
\Delta v_k - R v_k = -v_{k-2}[a_2 - \frac{(K_2 - p_{2,k-2})v_{k-2}}{2M_2} + c_2 u_k \\
\qquad -b_2 v_{k-2}] - R v_{k-2} \\
\Delta p_{1,k} - R p_{1,k} = -[a_1 p_{1,k-2} + \frac{(K_1 - p_{1,k-2})^2 u_k}{2M_1} & \text{in } \Omega \\
\qquad -(2b_1 u_{k-1} + c_1 v_{k-1})p_{1,k-2} + c_2 v_k p_{2,k-2}] \\
\qquad -R p_{1,k-2} \\
\Delta p_{2,k} - R p_{2,k} = -[a_2 p_{2,k-2} + \frac{(K_2 - p_{2,k-2})^2 v_k}{2M_2} \\
\qquad -(2b_2 v_{k-1} - c_2 u_k)p_{2,k-2} - c_1 u_{k-1} p_{1,k-1}] \\
\qquad -R p_{2,k-2}
\end{cases}
$$

where $\frac{\partial u_k}{\partial \nu} = \frac{\partial v_k}{\partial \nu} = \frac{\partial p_{1,k}}{\partial \nu} = \frac{\partial p_{2,k}}{\partial \nu} = 0$ on $\partial\Omega$. Here $R > 0$ is a sufficiently large constant. It is shown in [4] that if we assume all the hypotheses of Theorem 2.2, then the sequence of functions $u_k(x), v_k(x), p_{1,k}(x), p_{2,k}(x)$ defined above, satisfy the order relation

$$\begin{aligned} &u_0 \le u_2 \le \ldots \le u_{2r} \ldots \le u_{2r-1} \le \ldots \le u_1 \le u_{-1} \\ (2.8) \qquad &v_0 \le v_2 \le \ldots \le v_{2r} \ldots \le v_{2r-1} \le \ldots \le v_1 \le v_{-1} \\ &p_{i,0} \le p_{i,2} \le \cdots \le p_{i,2r} \cdots \le p_{i,2r-1} \le \cdots \le p_{i,1} \le p_{i,-1}, i = 1,2 \end{aligned}$$

for all $x \in \Omega$. Moreover, any solution (u, v, p_1, p_2) of problem (2.5) with the property (2.6) must satisfy

$$u_{2r} \le u \le u_{2r-1}, \quad v_{2r} \le v \le v_{2r-1}, \quad p_{i,2r} \le p_i \le p_{i,2r-1}, i = 1,2$$

for all positive integers r.

3 Optimal Control of Time-Periodic Competing-Species Diffusive Volterra-Lotka Systems

Recent results have also been obtained for the control of competing populations. Moreover, instead of the steady-state problem as in the last two sections, we consider time-periodic parabolic systems. More specifically, we consider T-periodic solution of the system

$$(3.1) \begin{cases} u_t - \Delta u - u[(a_1(x,t) - f_1(x,t)) - b_1 u - c_1 v] = 0 \text{ in } \Omega \times [0,\infty) \\ v_t - \Delta v - v[(a_2(x,t) - f_2(x,t)) - c_2 u - b_2 v] = 0 \text{ in } \Omega \times [0,\infty) \\ \frac{\partial u}{\partial \nu} = \frac{\partial v}{\partial \nu} = 0 \text{ on } \partial\Omega \times [0,\infty) \\ u(x,0) = u(x,T) \text{ and } v(x,0) = v(x,T) \text{ for } x \in \Omega \end{cases}$$

Here $a_i(x,t), f_i(x,t)$, $i = 1,2$ are functions in

$$L_+^\infty(\Omega \times [0,\infty)) = \{g \in L^\infty(\Omega \times [0,\infty)) | g \ge 0\},$$

with the property of being periodic in t with period T for $(x,t) \in \Omega \times [0,\infty)$. The parameters b_i, c_i, $i = 1,2$ are positive constants. For any constant vector $\vec{\delta} = (\delta_1, \delta_2), \delta_i > 0$, $i = 1,2$, we let

$$\begin{aligned} B_{\vec{\delta},T} = &\{(f_1, f_2) | f_i \in L_+^\infty(\Omega \times [0,\infty)), \text{ periodic in } t \text{ with period } T \\ &\text{and } f_i \le \delta_i, \ i = 1,2\}. \end{aligned}$$

For any $(f_1, f_2) \in B_{\bar{\delta},T}$, we define $(u,v) = (u(f_1,f_2), v(f_1,f_2))$ as a solution of problem (3.1). It will be uniquely defined under appropriate conditions. We define the payoff functional

$$\begin{aligned} J(f_1, f_2 &= \int_{\Omega\times[0,T)} [K_1 u(f_1,f_2) f_1 + K_2 v(f_1,f_2) f_2 \\ &\quad -M_1 f_1^2 - M_2 f_2] dx dt \end{aligned}$$

where K_i, M_i are constants. The problem is to find the periodic control $(f_1^*, f_2^*) \in B_{\bar{\delta},T}$ such that

$$J(f_1^*, f_2^*) = \sup_{(f_1,f_2)\in B_{\bar{\delta},T}} J(f_1, f_2). \tag{3.2}$$

For notational convenience, we denote $G = \Omega \times [0,T)$. Under appropriate hypotheses on $\delta_i, a_i, b_i, c_i, M_i$ and K_i the optimal control is found in [2] by finding periodic solution of the optimality system

$$\begin{cases} u_t - \Delta u - a_1 u + (b_1 + \frac{K_1 - z}{2M_1})u^2 + c_1 uv = 0 \text{ in } G \\ v_t - \Delta v - a_2 v + (b_2 + \frac{K_2 - w}{2M_2})v^2 + c_2 uv = 0 \text{ in } G \\ z_t + \Delta z - [2b_1 u - a_1 + c_1 v]z + \frac{(K_1 - z)^2}{2M_1} u - c_2 v w = 0 \text{ in } G \\ w_t + \Delta w - [2b_2 v - a_2 + c_2 u]w + \frac{(K_2 - w)^2}{2M_2} v - c_1 u z = 0 \text{ in } G \end{cases} \tag{3.3}$$

and then setting $f_1^* = \frac{u(x,t)}{2M_1}(K_1 - z), f_2^* = \frac{v(x,t)}{2M_2}(K_2 - w)$.

References

[1] F. He, A. Leung and S. Stojanovic, Periodic optimal control for parabolic Volterra-Lotka type equations, submitted.

[2] F. He, A. Leung and S. Stojanovic, Periodic optimal control for parabolic Volterra-Lotka type systems, to appear in J. of Computational and Appl. Math.

[3] A. Leung, Systems of Nonlinear Partial Differential Equations, Applications to Biology and Engineering, Kluwer Academic Publishers, Dordrecht/Boston, 1989.

[4] A. Leung, Optimal harvesting-coefficient control of steady-state prey-predator diffusive Volterra-Lotka systems, submitted.

[5] A. Leung and S. Stojanovic, Optimal control for elliptic Volterra-Lotka type equations, to appear in J. Math. Anal. Appl.

Third International Colloquium on Differential Equations pp. 157-168 (1993)
D. Bainov and V. Covachev (Eds)

Stokes Multipliers For A Certain n-th ORDER DIFFERENTIAL EQUATION

T.K.PUTTASWAMY
Department of Mathematical Sciences, Ball State University
Muncie, Indiana–47306, U.S.A.

ABSTRACT

This paper is devoted to the solution in the large of the differential equation

$$z^n(\mu^2 - 2\mu z^m + z^{2m})\frac{d^n y}{dz^n} + \sum_{k=0}^{i}(a_k + b_k z^m + c_k z^{2m})\frac{d^k y}{dz^k} = 0 \qquad 0 \leq i < n-1$$

where m is an arbitrary positive integer, and the variable z and the constants $\mu, a_k, b_k, c_k (k = 0, 1, 2, \ldots, i)$ are complex with $\mu \neq 0$. We shall also assume that the difference of no two roots of the indicial equation about $z = 0$ is congruent to zero modulo m.

Key Words

Singular Points,Asymetric series, Stokes Multipliers and Factorial Series.

1 INTRODUCTION

Analytic Theory of linear differential equations in the complex plane has a local as well as a global aspect. The local part of the theory is fairly developed, where as global results, are sparse and fragmentary. The global results have been carried out mostly in the case of special functions, which satisfy second order linear differential equations. A typical global problem is the so called connection problem, the problem of decomposing a solution at a given point into a linear combination of fundamental solutions at some other point. To be specific, if z_1 and z_2 are two distinct points and if p(z) is a solution at z_1, how can we express p(z) as a linear combination of linearly independent solutions at z_2 ? If z_1 and z_2 are ordinary points, the problem is simple.All we have to do is to carry the analyutic continuation. If z_1 or z_2 or both are singular points ,then the story is totally different.Even for seemingly simple looking differential equations, the problem becomes quite complicated. In particular, if z_1 or z_2 or both are irregular singular points, the degree of difficulty becomes multifold, for ,the neighbourhood of an irregular singular point is not a homogeneous media with reference to the behavior of solutions.In fact ,the neighbourhood breaks up into complimentary regions, usually sectorial in shape and in each of which the solutions will have a different asymptotic development. As a general rule, the degree of difficulty increases with the number of singular points, the differential equation has.

In order to introduce the investigations of this paper, let us take for consideration the linear homogenious differential equation of n th order

$$z^n(\mu^2 - 2\mu z^m + z^{2m})\frac{d^n y}{dz^n} + \sum_{k=0}^{i}(a_k + b_k z^m + c_k z^{2m})z^k\frac{d^k y}{dz^k} = 0$$

$$0 \leq i < n-1 \tag{1.1}$$

Here m is an arbitary positive integer. The variable z and the constants $\mu, a_k, b_k, c_k (k = 0, 1, 2, \ldots, i)$ are complex with$\mu \neq 0$.If $\mu_1, \mu_2, \mu_3, \ldots, \mu_m$ are the roots of $\mu - z^m = 0$ (1.2)

then according to Fuch's theory, our differential equation (1.1) will have regular singular points at z=0, z=$\mu_1, \mu_2, \ldots, \mu_m$ and at $z = \infty$. In all, (1.1) will have $m+2$ regular singular points. Since m is an arbitrary positive integer, the differential equation (1.1) will have arbitrary number of regular singular points.

If

$$\{h\}_k = h(h-1)(h-2)\ldots(h-k+1) = \prod_{j=0}^{k-1}(h-j), \qquad k \geq 1$$

and $\{h\}_0 = 1$

Then the indicial equation ablout z=0 is found to be

$$\mu^2\{h\}_n + \sum_{k=0}^{i} a_k\{h\}_k = 0 \tag{1.3}$$

We shall also assume that the roots h_1,h_2,...,h_n of (1.3) are such that the difference of no two of them is congruent to zero m. Then, the existing theory of linear differential equations assures that (1.1) will have n linearly independent solutions $y_j(z)$, j=1,2,...,n about z=0 of the form

$$y_j(z) = z^{h_j}\sum_{k=0}^{\infty} g_j(k)z^{mk}$$

$$j = 0, 1, 2, \ldots, n$$

with $g_j(0) = 1$

(1.4)

Here, each $g_j(k), k = 1, 2, 3, \ldots$ is a determinate function of k and each of these series solutions in (1.4) will converge in a punatured circle drawn about $z=0$ and extending up to the nearest of singular points $z = \mu_i (i = 1, 2, \ldots, m)$ outside this circle, each becomes divergent. The behaviour of $y_j(z)(j = 1, 2, \ldots, n)$ in the neighbourhood of $z = \infty$ is not available, as far as Fuch's theory is concerned, even though $z = \infty$ is a regular singular point. However, we know from the existing theory of linear differential equations, each of the solutions about $y_j(z)$,(j=1,2,...,n) may be expressed as some linear combination of n linearly independent solutions about $z = \infty$ having, in general, the forms

$$\tilde{y}_j(z) = \frac{1}{z^{l_j}} \sum_{k=0}^{\infty} \frac{\tilde{g}_j(k)}{z^k}, \tilde{y}_j(0) = 1 \; for \qquad \tilde{y} = 1, 2, \ldots, n \tag{1.5}$$

where $l_j (j = 1, 2, \ldots, n)$ are the identical exponents about $z = \infty$ and where each of the power series solutions in (1.5) is convergent, when $| z |$ is sufficiently large. The determination of the precise nature of such a linear relationship, including the connection coefficients, is a problem of extreme difficulty. These connection coefficients are called Stokes Multipliers. until a convergent method of computing these Stokes Multipliers is devised, a differential equation can not be considered as solved in the large. If is to the solution of this problem in the large that this paper has been addressed.

MAIN RESULTS AND THE MAIN THEOREM

2 Solutions $y_i(z)(i = 1, 2, \ldots, n)$ $\quad$ $about z = 0$

We begin by determining the recurrence relations satisfied by each g_z in (1.3) These recursive relations will be sencond order difference equations of the form

$$p_2(k)g(k+2) + p_1(k)g(k+1) + p_0 g(k) = 0 \tag{2.1}$$

where

$$p_2(k) = \mu^2 \{km + h + 2m\}_n + \sum_{j=0}^{i} a_j \{km + h + 2m\}_j \tag{2.2}$$

$$p_1(k) = -2\mu\{km + h + m\}_n + \sum_{j=0}^{i} \{km + h + m\}_j \tag{2.3}$$

$$p_0(k) = \{km + h\}_n + \sum_{j=0}^{i} c_j \{km + h\}_j \tag{2.4}$$

We observe that $g(k)$ for $k = 1, 2, \ldots$. are completely and correctly determined by (2.1) if we arbitrarly take $g(-1) = 0$ as we do. Furthermore, we make the assumption that when one of $h_j (j = 1, 2, \ldots, n)$ is used for h, no root of $p_0(k-2) = 0$ is zero or a negative integer, in which case, we will have$g(-k) = 0$ for $k = 2, 3, \ldots$ Since we wish to regard k as a complex variable, to emphasize this, we replace k by s.

Our first objective, then, is to determine that analytic function $u(s)$ of the complex variable s, which is a solution of (2.1) and which takes on initial values $u(0) = 1$ and $u(-1) = 0$. To do this, we let

$$U_1(s) = u(s) \qquad (2.5)$$

$$U_2(s) = u(s+1) \qquad (2.6)$$

Then (2.1) can be put in the matrix form

$$G(s+1) = A(s)G(s) \qquad (2.7)$$

where $G(s)$ is a 2x2 matrix which may be decomposed into two column vectors,each with two components and $A(s)$ is a 2x2 matrix which can be represented as a power series in $\frac{1}{s}$ and is convergent in some neighbourhood of $s = \infty$.

Therfore there is a finite s_0 such that the series

$$A(s) = A_0 + \frac{A_1}{s} + \frac{A_2}{s^2} + \ldots. \qquad (2.8)$$

converges for $| s | > s_0$.
Here,

$$A_0 = \begin{pmatrix} 0 & 1 \\ \gamma_0 & \omega_0 \end{pmatrix}$$

$$A_j = \begin{pmatrix} 0 & 0 \\ \gamma_j & \omega_j \end{pmatrix}, \qquad j=1,2,\ldots,n$$

with

$$\gamma_0 = -\frac{1}{\mu^2}$$

$$\omega_0 = \frac{2}{\mu}$$

$$\gamma_1 = \frac{n}{\mu^2}$$

$$\omega_1 = -\frac{2n}{\mu}$$

(2.9)

By using a paper by W.J.A.Culmer and W.A.Harris Jr [1], we find that the two linearly independent solutions $u_1(s)$ and u_2 of (2.1) are given by known factorial series as follows:

$$u_1(s) = (\frac{1}{\mu})^s s^{r_1}[1 + E_1(s)] \quad (2.11)$$

$$u_2(s) = (\frac{1}{\mu})^s s^{r_2}[1 + E_2(s)] \quad (2.12)$$

where

$$r_1 = \frac{p + \sqrt{p^2 + 4\mu q}}{2} \quad (2.12)$$

$$r_2 = \frac{p - \sqrt{p^2 + 4\mu q}}{2} \quad (2.13)$$

with

$$p = \mu w_1 + 1 \quad (2.14)$$

$$q = \mu r_2 + 1 \quad (2.15)$$

$u_1(s)$ and $u_2(s)$ are analytic at every point in some right half plane and in this right half plane $E_j(s) \to 0$ as $s \to \infty$ for $j = 1, 2, \ldots$.

By rendering precise meaning through certain conventions and by analytic continuation, we find that $u_1(s)$ and $u_2(s)$ are single valued and analytic throughout the s-plane except for poles at the points.

$s = \alpha_j - k, \quad k = 0, 1, 2, \ldots$ (2.16) where $\alpha_j (j = 1, 2, \ldots, n)$ are the roots of $P_0(s) = 0$

Next , we determine the constant a and b such that $u(s) = au_1(s) + bu_2(s)$ (2.17)

The constants a and b are evidently to be determined from the initial conditions $u(0) = 1$ and $u(-1) = 0$ However, in order to obtain such a function $u(s)$, we shall assume that none of the poles (2.16) is an integer. the result of removing this restriction, we shall examine at a later point, but it may be observed that in part this restriction was made at the outset, when we assumed that none of the roots $P_0(s-2) = 0$ is zero or a negative integer.

If we put

$$D(s) = \begin{vmatrix} u_1(s) & u_2(s) \\ u_1(s+1) & u_2(s+1) \end{vmatrix} \tag{2.18}$$

We find that

$$a = -\frac{u_2(-1)}{D(-1)} \quad \text{and} \quad b = \frac{u_1(-1)}{D(-1)} \tag{2.19}$$

Our next objective is to find the functional value of the Casorati's determinant $D(s)$.

By Heyman's theorem,$D(s)$ satisfies

$$\frac{D(s+1)}{D(s)} = \frac{P_0(s)}{P_2(s)} = \frac{\prod_{k=1}^{n}(s-\alpha_k)}{\mu^2\prod_{k=1}^{n}(s-\beta_k+2)} \tag{2.20}$$

where $\beta_j, j = 1,2,\ldots,n$ are the roots of $P_2(s-2) = 0$

We observe that, in view of (1.3), 0 is a root of $P_2(s-2) = 0$ let β_n be this root.

Then,

$$D(s) = \frac{k^*(\frac{1}{\mu^2})^s\prod_{k=1}^{n}(s-\alpha_k)}{\Gamma(s+2)\prod_{k=1}^{n-1}\Gamma(s-\beta_k+2)} \tag{2.21}$$

where

$$k^* = \begin{cases} -\mu\sqrt{p^2+4q} & \text{if } p^2+4q \neq 0 \\ -\mu^3 & \text{if } p^2+4q = 0 \end{cases} \tag{2.21}$$

So

$$D(-1) = \frac{k^*\mu^2\prod_{k=1}^{n}(-1-\alpha_k)}{\prod_{k=1}^{n-1}\Gamma(1-\beta_k)} \tag{2.23}$$

Here, it is observed that in passing when $h = h_j (j =, 1, 2.\ldots, n-1)$ none of the expressions $1-\beta_j$ $(j =, 1, 2, \ldots, n-1)$ can be zero or a negative integer. Hence$D(-1) \neq 0$.For,if $1-\beta_k = N$,where $N = 0,1,2,\ldots$ we will have $\beta_j = N$,where $N = 1,2,\ldots$ .Since $\beta_j (j = 1,2,\ldots,n-1)$ are the roots of $P_0(s-2) = 0$ this would imply that the difference of two roots of the indecial equations (1.3) is congruent to zero modulo m, which contradicts our hypothesis.

Again, we note that the numerator in (2.23) is necessarily finite, in as much as we have assume that $\alpha_j, j = 1,2,\ldots,n$ are non integers.

So in view of (2.17) and (2.19),

$$u(s) = \frac{1}{k^*\mu^2}\frac{\prod_{k=1}^{n}(1-\beta_k)}{\prod_{k=1}^{n}\Gamma(1-\alpha_k)}V(s) \tag{2.24}$$

where

$$V(s) = \begin{vmatrix} u_1(-1) & u_2(-1) \\ u_1(s) & u_2(s) \end{vmatrix} \tag{2.25}$$

Next, we prove the following lemma:

lemma

If $l_j(j = 1, 2, \ldots, n)$ are the indicial exponents about the regular singular point $z = \infty$ for differential equation (1.1)

then, $\alpha_j = -(\frac{l_j+h}{m})$ for $j = 1, 2, \ldots, n$ (2.26)

If follows that each of the solutions $y_j(z)$ of (1.4) about the regular singular $z = 0$ for our differential equation (1.1) is of the form

$$y(z) = z^h[\frac{-u_2(-1)}{D(-1)} \sum_{k=0}^{\infty} G_1(k)(\frac{z^m}{\mu})^s + \frac{u_1(-1)}{D(-1)} \sum_{k=0}^{\infty} G_2(k)(\frac{z^m}{\mu})^k] \tag{2.27}$$

where $G_1(s) = s^{r_1}[1 + E_1(s)]$ (2.28)

$$G_2(s) = s^{r_2}[1 + E_2(s)] \tag{2.29}$$

3 Asymptotic Behaviour of the solutions $y_i(z)(i = 1, 2, \ldots, n)$

We apply the modified version of W.B.Ford's Theorem [2] to determine the asymptotic developments of the functions, defined by the two series in (2.27). Then, we find that the fundamental solution $y(z)$ about $z = 0$ as defined by (1.4) where $| z |$ is large and when z lies along any ray except for which $\arg z^m = \arg \mu$ may be expanded asymptotically in the form

$$y(z) \sim -z^h R(z) \tag{3.1}$$

where $R(z)$ represents the sum of the residues of the function

$$\frac{\pi u(s)(-z)^s}{\sin \pi s} \tag{3.2}$$

at the poles (2.26).

Our next problem then is find $R(z)$. If $R_{jk}(k)$ is the residue of (3.2) at the pole $s = \alpha_j - k = \frac{-h-l_j}{m} - k$, then

$$R(z) = \sum_{i=1}^{n} \sum_{k=0}^{\infty} R_{jk}(z) \tag{3.3}$$

At this stage, we shall also assume that none of $l_j - l_k$ $(j \neq k, j, k = 1, 2, \ldots, n)$ is an integer. Then our poles (2.26) will all be simple poles of (3.2).
We define,

$$f_0(h) = \mu^2\{k\}_n + \sum_{j=0}^{i} a_j\{k\}_j \tag{3.4}$$

$$f_1(h) = -2\mu\{k\}_n + \sum_{j=0}^{i} b_j\{k\}_j \tag{3.5}$$

Then, we observe, that in view of (1.3), $f_0(h) = 0$. Moreover, from (2.2) and (2.3), we have

$$P_2(s) = f_0[m(s+2)+h] \tag{3.6}$$

and

$$P_1(s) = f_1[m(s+1)+h] \tag{3.7}$$

After computing $R_{jk}(z)$ we find from (3.3) that when $| z |$ is large and when z lies along any ray except $\arg z^m = \arg \mu$,

$$y(z) \sim \sum_{k=1}^{n} c_k(h, l_1, l_2, \ldots, l_n)\frac{1}{z^{L_k}}[1 + \frac{(\ )}{z^m} + \frac{(\ )}{z^{2m}} + \ldots] \tag{3.8}$$

where

$$c_1(h, l_1, l_2, \ldots, l_n) = \frac{e^{\pi i(h+l_1)}\Gamma(2 - \frac{h+l_1}{m})\pi_{k=1}^{n-1}\Gamma(1-\beta k)}{\mu^2 k^* \pi_{k=2}^{n}(l_k - l_1)\pi_{k=2}^{n}\Gamma(\frac{h+l_k}{m} - 1)} W(-h - l_1) \tag{3.9}$$

with $W(s) = V(s+2)f_0(2m - l_1) + V(s+1)f_1(m - l_1)$ (3.10)
and $c_k(h, l_1, l_2, \ldots, l_n)$ for $k = 2, 3, \ldots, n$ is obtained from (3.9) by interchanging l_1 and l_k.

In arriving at this result, we have placed several restrictions. In fact, we have assumed that none of the quantities $\alpha_k = -\frac{h+l_k}{m}(k = 1, 2, \ldots, n)$ and $\alpha_k - \alpha_j = \frac{l_j - l_k}{m}$ $(k \neq j, j, k =, 1, 2 \ldots, n)$ an integer.

Assuming that $\alpha_k - \alpha j = \frac{l_j - l_k}{m}(k \neq j, j, k = 1, 2, \ldots, n)$is not an integer, which means that the difference of no two indicial exponents $l_j$and $l_k (k \neq j, j, k = 1, 2, \ldots, n)$ is congruent the zero modulo m, we shall now examine the effects upon (3.9), when one of $\frac{h+l_k}{m}, k = 1, 2, 3, \ldots, n$ is an integer. In this case, it is to be noted that in no case two of these quantities is an integer, for, this would imply that the difference of two identical exponents about $z = \infty$ is congruent to zero modulo m, contradicting the hypothesis. Moreover, it will suffice to consider nearly the case in which $\frac{h+l_1}{m}$ is an integer in as much as all possible cases may then covered by merely interchanging those of l_1 and $l_j(j = 2, 3, \dots, n)$

We shall suppose that $\frac{h+l_1}{m}$ is an integer. Let us devide the possible integral values of $\frac{h+l_1}{m}$ in two classes

(a) $\frac{h+l_1}{m}$ is on integer < 2

(b) $\frac{h+l_1}{m}$ is an integer ≥ 2

Case (a) $\frac{h+l_1}{m}$ **is an integer** < 2

If we now form the differential equation in which μ^2 in 1.1 is changed to $\mu^2 - E$, where E is arbitrary small and positive, the resulting values of $l_j(j = 1, 2, \dots, n)$ remains as before, in as much as they depend only $c_j, j = 0, 1, 2, \dots, i$. The resulting value of h will be of the form $h - \eta$, where η becomes infinitely small with E and atleast after E has been taken sufficiently small, none of the quantities $\frac{h-l_k-\eta}{m}$ is an integer. The solution $y(z)$ corresponding to (1.4) of the new differential equation will therefore, be developable asymptotically in the form obtained by (3.8) replacing $\frac{h}{m}$ by $\frac{h-\eta}{m}$ and the resulting $c_k(h-\eta, l_1, l_2, \dots, l_n)$ determined from (3.9).

After the expressions thus obtained have been examined, it is noted that when $\frac{h+l_1}{m}$ is an integer < 2, each of the following limit exists:

$$\lim_{\eta \to 0} \frac{\Gamma(2 - \frac{h+l_k-\eta}{m})}{\prod_{j=1\, j \neq k}^{n} \Gamma(\frac{h+l_j-\eta}{m} - 1)} \qquad \texttt{for } \mathtt{k} = 1, 2, \dots, \mathtt{n} \tag{3.10}$$

Hence we conclude that so long as $\frac{h+l_1}{m}$ is an integer belonging to class (a) mentioned above the relation (3.8) continues as before.

Case (b) $\frac{h+l_1}{m}$ **is an integer** ≥ 2

First we consider the case $\frac{h+l_1}{m} = 2$ Proceeding as before, we obtain the new differential equation obtained from (1.1) by replacing μ^2 by $\mu^2 - E$ and hence h by $h - \eta$ in (3.8), we endevour to examine the limit as $\eta \to 0$ of $c_1(h - \eta, l_1, l_2, \dots, l_n)$ as determined by (3.9). In doing so, there is involved, aside from the factor which clearly approaches a limit, the product

$$\Gamma(\frac{\eta}{m})[f_0(h)V\frac{\eta}{m} + f_1(h-m)V(\frac{\eta}{m} - 1)] \tag{3.11}$$

since $f_0(h) \equiv 0$, we have $f_0(h)V(\frac{\eta}{m}) = 0$

Hence limits of (3.11) as $\eta \to 0$

$$= \lim_{\eta \to 0} \Gamma(\frac{\eta}{m}) f_1(h-m) V(\frac{\eta}{m} - 1) \qquad \textbf{(3.12)}$$

Hence, as $\eta \to 0$, $\Gamma(\frac{\eta}{m})$ becomes infinite to the first order.

Also,

$$\lim_{\eta \to 0} f_1(h-m) V(\frac{\eta}{m} - 1)$$

$$= f_1(h-m) V(-1) = 0$$

Thus the limit of (3.12) as $\eta \to 0$ is indeterminate. We can however evaluate it upon replacing

$$\Gamma(\frac{\eta}{m}) = \frac{\Gamma(1 + \frac{\eta}{m})}{\frac{\eta}{m}}$$

The result being $f_1(h-m)[\frac{d}{d\eta} V(\frac{\eta}{m} - 1)]_{\eta=0}$

$$= m f_1(h-m)[\frac{d}{ds} V(\frac{s}{m} - 1)]_{s=0}$$

$$= f_1(h-m)[\frac{d}{ds} V(s-1)]_{s=0} \qquad \textbf{(3.13)}$$

or, as we shall prefer to write it for future purposes,

$$[f_0(2m - l_1)\frac{d}{ds} V(s+2) + f_1(m - l_1)\frac{d}{ds} V(s+1)]_{s=-\frac{h+l_1}{m}=-2} \qquad \textbf{(3.14)}$$

Let us suppose next that $\frac{h+l_1}{m}$ = an integer $N > 2$ .1 cm Then, reasoning as in the case $\frac{h+l_1}{m} = 2$, we replace h in (3.9) by $h - \eta$ and attempt to obtain the limit of the resulting expressions as $\eta \to 0$. No difficulties are encountered except in connection with the product

$$\Gamma(2 - N + \frac{\eta}{m})[f_0(2m - l_1)V(2 - N + \frac{\eta}{m}) + f_1(m - l_1)V(1 - N + \frac{\eta}{m})] \qquad \textbf{(3.15)}$$

But

$$\Gamma(2 - N + \frac{\eta}{m}) = \frac{(-1)^N \Gamma(1 + \frac{\eta}{m})}{(N - 2 - \frac{\eta}{m})(n - 3 - \frac{\eta}{m}) \dots (1 - \frac{\eta}{m})(\frac{\eta}{m})}$$

since $N > 2$

$$\lim_{\eta \to 0} V(2 - N + \frac{\eta}{m}) = V(2 - N) = 0$$

and

$$\lim_{\eta \to 0} V(1 - N + \frac{\eta}{m}) = V(1 - N) = 0$$

Hence, we see the limits of 3.15 as $\eta \to 0$ is

$$\frac{(-1)^N}{\Gamma(N-1)}[f_0(2m-l_1)\frac{d}{ds}V(s+2)+f_1(m-l_1)\frac{d}{ds}V(s+1)]_{s=0} \tag{3.16}$$

Finally we note that (3.14) may be obtained from (3.16) by putting $N = 2$, so that when $\frac{h+l_1}{m}$ is an integer ≥ 2 , we conclude that relation (3.8) for $y(z)$ in which $c_1(h, l_1, l_2, \ldots, l_n)$ is no longer to be determined by (3.9) but by the following:
$c_1(h, l_1, l_2, \ldots, l_n)$

$$= \frac{\prod_{k=1}^{n-1}\Gamma(1-\beta_k)[f_0(2m-l_1)\frac{d}{ds}V(s+2)+f_1(m-l_1)\frac{d}{ds}V(s+1)]}{\mu^{2k*}\Gamma(N-1)\prod_{k=2}^{n}(l_k-l_1)\prod_{j=2}^{n}\Gamma(\frac{h+k_j}{m}-1)}_{s=-N}$$

$$(3.17)$$

while $c_j(h, l_1, l_2, \ldots, l_n)$ for $j = 2, 3, \ldots, n$ are to be determined as before by merely interchanging l_1 and l_j in (3.9).

In summary, we arrive at the following main theorem:

Theorem:

If the indicial exponents $l_j (j = 1, 2, \ldots, n)$ of the regular singular point $z = \infty$ of the differential equation (1.1) are such that the difference of no two of them is congruent to zero modulo m, then the solution $y_j(z), j = 1, 2, \ldots, n$ about the regular singular point $z = 0$ as defined in (1.4), when considered for values of z of large modulus and for which $\arg z^m \neq \arg\mu$ may be developed asymptotically in the form
$y_j(z) \sim \sum_{k=1}^{n} c_k(h, l_1, l_2, \ldots, l_n)\tilde{y}_k(z)$ where $\tilde{y}_k(z), k = 1, 2, \ldots, n$ are the solutions (1.5)

(a) In case, none of the quantities $\frac{h+l_k}{m}, k = 1, 2, \ldots, n$ is an integer or in case $\frac{h+l_1}{m}$ is an integer < 2 , the constants $c_1(h_j, l_1, l_2, \ldots, l_n)$ is determined by (3.9) in which h_j is used for h , while the remaining constants $c_j(h_j, l_1, l_2, \ldots, l_n)$ ar determined by interchanging l_1 and l_j.

(b) In case, $\frac{h+l_1}{m} = N =$ an integer ≥ 2, then constant $c_1(h_j, l_1, l_2, \ldots, l_n)$ is determined by the use of $h = h_j$ in (3.17), while the constants $c_j(h_j, k_1, k_2, \ldots, k_n)$ continues to be determined from (3.9) as in case (a).

In arriving at the above theorem, it must be understood that we have assumed $\mu \neq 0$ in (1.1) and the indicial exponents $h_j (j = 1, 2, \ldots, n)$ are such that the difference of no two of them is congruent to zero modulo m. Moreover, in as much as the series is $\tilde{y}_j(z), j = 1, 2, \ldots, n$ are known in the present instance to be convergent for the indicated values of z, the symbol $\sim$ as used above may be changed to $=$ and the resulting equation may be regarded as those which join the fundeamental solutions (1.4) with the fundamental solutions (1.5) upon a suitably constructed Riemann surface having cut the line drawn from $z = \mu$ to $z = \infty$

Due to the limitation in space, the case when the difference of the indicial exponants about $z = \infty$ is congruent to zero modulo m will not be discussed here and will be published elsewhere as a sequel to this paper. This paper extends and generalizes [3] and [4].

REFERENCES

[1] W.J.A.Culmer and W.A.Harris Jr, Convergent Solutions of ordinary Linear Homogeneous Difference Equations,Pac.Jour.of Math. 13 (1963),1111-1138.

[2] W.B.Ford, Asymptotic Developments of Functions Defined by Maclaurin Series, Chelsa publiching company (1960).

[3] T.K.Puttaswamy, Solution in the Large of a Certain n^{th} Order Differential Equation, Proceedings of Ramanujan Birth Centenary Year International symposium on Analysis, Macmillan India Limited, 1989, 451-465.

[4] T.K.Puttaswamy, A connection problem for a certain n-th order Homogeneous Differential Equation, Proceeding of the International Conference on Theory and Applications of Differential Equations, Ohio University Press, 1989, 317-321.

[5] T.K.Puttaswamy, A generalization of a theorem of W.B.Ford, Proceedings of the International Conference on New Trends in Geometric Function Theory and Applications, World Scientific, 1991, 90-95.

Third International Colloquium on Differential Equations pp. 169-175 (1993)
D. Bainov and V. Covachev (Eds)

SUPERSYMMETRIC QUANTUM MECHANICS AND MULTI-SOLITON SOLUTIONS OF HIGHER ORDER K-dV EQUATIONS

C. V. SUKUMAR
Theoretical physics, University of Oxford
1 Keble road, Oxford OX1 3NP, England

ABSTRACT: It is shown that it is possible to construct higher order equations of the K-dV type for which explicit expressions for the n-soliton solution may be given. For these equations an n-soliton solution and an $(n+1)$- soliton solution may be paired as the fermionic and bosonic components of a supersymmetric system.

Keywords: Supersymmetry, Quantum Mechanics, Solitons, K-dV Equation.

In field theory, supersymmetry is a symmetry that generates transformations between bosons and fermions. Starting from a supersymmetric field theory, it is possible to constuct supersymmetric Quantum Mechanics (Witten 1981). A supersymmetric Hamiltonian may be defined in terms of charges that obey the same algebra as that of the generators of supersymmetry in field theory (Freedman and Cooper 1983). It has been shown (Andrianov *et al* 1984, Sukumar 1985) that every one dimensional Hamiltonian H_1 can have a partner H_2 such that H_1 and H_2 taken together may be viewed as the components of a supersymmetric Hamiltonian. The consequences of supersymmetry for the spectra of H_1 and H_2 have been extensively discussed. The supersymmetric pairing may be used to generate a H_2 by eliminating the groundstate of H_1 or adding a state below the groundstate of H_1 or maintaining the spectrum to be the same as that of H_1. The standard inverse scattering method based on the Gelfand-Levitan procedure for finding a new potential by (1) eliminating the groundstate of a given potential, (2) adding a state below the groundstate of a given potential and (3) generating the phase equivalent family of a given potential may each be shown to consist of the successive application of two appropriately chosen supersymmetric transformations (Sukumar 1985).

The connection between the algebra of supersymmetry and the inverse scattering method may be used to construct one dimensional potentials with any number of boundstates at any chosen energies. The reflection coefficient of the potential so constructed is related to the reflection coefficient of a reference potential which supports no boundstates. All potentials connected by supersymmetry to a reference potential that is reflectionless therefore have vanishing reflection coefficient . It has been shown that using this idea a reflectionless potential with any number n of non-degenerate boundstates at arbitrarily chosen energies may be constructed (Sukumar 1986). Choosing the reduced mass $\mu = 1/2$, the potential with boundstates at energies E_j , $j = 1, 2, ...n$ may be represented in the form

$$\begin{aligned} V_n &= -2\frac{d^2}{dx^2} \ln Det\ B \ , \\ B_{ij} &= \frac{d^{i-1}}{dx^{i-1}} \phi_j(x, \Gamma_j) \ , \ i, j = 1, 2, ...n \ , \\ \phi_j &= (1/2) \left[e^{\Gamma_j x + \theta_j} + (-)^{j+1} e^{-\Gamma_j x - \theta_j} \right] \ , \\ E_j &= -\Gamma_j^2 \ , \ \Gamma_n > \Gamma_{n-1} > .... > \Gamma_1 \ , \end{aligned} \tag{1}$$

where θ_j are arbitrary phase factors. It must be noted that for odd values of j, ϕ_j is a *cosh* function and hence nodeless while for even values of j, ϕ_j is a *sinh* function with a single node. Such a structure for ϕ_j ensures that V_n is free of singularities. In terms of the normalised eigenstates of V_n given by

$$\Psi_n(x, E_j) = (B^{-1})_{jn} \ , \ j = 1, 2, ...n \tag{2}$$

the potential V_n may be written as

$$\begin{aligned} V_n &= -2 \sum_{j=1}^{n} \alpha_j \ \Psi_n^2(E_j) \\ &\text{where} \\ \alpha_j &= \prod_{k \neq j} |\Gamma_j^2 - \Gamma_k^2| \ \Gamma_j^2 \ . \end{aligned} \tag{3}$$

A time dependence for V_n can now be introduced by allowing the phases θ_j to have a time dependence. Connection to the solutions of the K-dV equation may be established by allowing the phases to vary according to

$$\theta_j(t) = \theta_j(0) - 4\Gamma_j^3 \ t \ . \tag{4}$$

The potential $V_n(x, t)$ given by equation (1) with θ_j replaced by $\theta_j(t)$ then supports n boundstates at energies E_j , $j = 1, 2, ..n$. This potential has been shown (Sukumar

1986) to be identical to the n-soliton solution of the K-dV equation

$$\left(\frac{d^3}{dx^3} - 6V_n \frac{d}{dx} + \frac{d}{dt} \right) V_n(x,t) = 0 \tag{5}$$

given by (Kay and Moses 1956, Gardner *et al* 1967, Scott *et al* 1973)

$$\begin{aligned} V_n(x,t) &= -2\frac{d^2}{dx^2} \ln Det\ M \ , \\ M_{ij} &= \delta_{ij} + \lambda_j(x)\, \lambda_k(x) \ , \\ \lambda_j(x) &= C_j\, e^{-\Gamma_j y_j} \ , \\ C_j^2 &= 2\,\Gamma_j \prod_{k \neq j} \frac{|\Gamma_k + \Gamma_j|}{|\Gamma_k - \Gamma_j|} \ , \\ y_j &= x - 4\,\Gamma_j^2\, t \ , \end{aligned} \tag{6}$$

in which δ_{ij} is the Kronecker delta function. It is possible to verify by direct calculation that V_n defined by equations (1) and(4) satifies the K-dV equation.

Discussion of supersymmetry in the context of soliton solutions of the K-dV equation is relevant (Kwang and Rossner 1985, Sukumar 1986) because the n-soliton potential V_n with energies determined from $(\Gamma_1, \Gamma_2, ...\Gamma_n)$ and the $(n+1)$-soliton potential V_{n+1} with energies determined from $(\Gamma_1, \Gamma_2, ...\Gamma_n, \Gamma_{n+1})$ can be related to the two partners of a supersymmetric Hamiltonian as shown below. The Hamiltonians corresponding to V_{n+1} and V_n may be written as

$$\begin{aligned} H_{n+1} &= -\frac{d^2}{dx^2} + V_{n+1} = A_n^-(E_{n+1})\, A_n^+(E_{n+1}) + E_{n+1}, \\ H_n &= -\frac{d^2}{dx^2} + V_n = A_n^+(E_{n+1})\, A_n^-(E_{n+1}) + E_{n+1}, \\ A_n^{\pm}(E_{n+1}) &= \left[\pm\frac{d}{dx} + \left(\frac{d}{dx} \ln \Psi_n(E_{n+1}) \right) \right] \end{aligned} \tag{7}$$

where $\Psi_n(E_{n+1})$ is a nodeless, non-normalisable solution of the Schrödinger equation for the potential V_n at energy E_{n+1} which lies below the groundstate of V_n at energy E_n. $(H_{n+1} - E_{n+1})$ and $(H_n - E_{n+1})$ are the diagonal elements of a supersymmetric Hamiltonian given by the anticommutator

$$H = \{Q, Q^{\dagger}\}$$

where

$$Q = \begin{pmatrix} 0 & 0 \\ A_n^-(E_{n+1}) & 0 \end{pmatrix} \tag{8}$$

and

$$Q^\dagger = \begin{pmatrix} 0 & A_n^+(E_{n+1}) \\ 0 & 0 \end{pmatrix}$$

The eigenstates of V_n and V_{n+1} are then connected by the transformations given by

$$\begin{aligned} \Psi_{n+1}(E_i) &\sim A_n^-(E_{n+1})\,\Psi_n(E_i)\ ,\ i=1,2,...n \\ &\text{and} \\ \Psi_{n+1}(E_{n+1}) &\sim 1/\Psi_n(E_{n+1})\ . \end{aligned} \tag{9}$$

Furthermore

$$V_{n+1} = V_n - 2\frac{d^2}{dx^2}\ln\Psi_n(E_{n+1})\ . \tag{10}$$

In this manner a potential V_{n+1} with an additional boundstate at energy E_{n+1} below the groundstate energy E_n of the potential V_n may be constructed using a solution of the Schrödinger equation for the potential V_n. It is known (Scott *et al* 1973) that the time evolution of the eigenstates of V_n is given by

$$\left(\frac{d}{dt} + 4\frac{d^3}{dx^3} - 6V_n\frac{d}{dx} - 3\frac{dV_n}{dx}\right)\Psi_n(E_i,x,t) = 0,\ i=1,2,...n\ . \tag{11}$$

V_n defined by equation (1) satisfies the K-dV equation (5) when the time evolution of the phase is given by equation (4). It is easy to see that this time evolution of the phase is tantamount to the requirement that the elements of the matrix B appearing in equation (1) satisfy

$$\frac{d}{dt}B_{ij} = -4\frac{d^3}{dx^3}B_{ij}\ . \tag{12}$$

The question then arises as to whether it is possible to find other non-linear equations whose solutions retain the structure obtained from arguments of supersymmetry and explicitly given by equation (1) but have a time dependence for θ_j different from that for the K-dV equation for which θ_j must satisfy equation (4). Imposition of a condition analogous to equation (12) but involving different derivatives of x may be attempted. This leads us to consider a condition of the form

$$\frac{d}{dt}B_{ij} = -2^{m-1}\frac{d^m}{dx^m}B_{ij}\ . \tag{13}$$

It was noted earlier that the ϕ_j appearing in equation (1) is either a *cosh* or a *sinh* function of x and t. Such a structure for ϕ_j restricts the m values in equation (13) to be odd integers. For $m = 1$, equation (13) can be satisfied if

$$\begin{aligned} \theta_j(t) &= \theta_j(0) - \Gamma_j t \quad and \\ \phi_j &= (1/2)\left[e^{\Gamma_j(x-t)+\theta_j(0)} + (-)^{j+1} e^{-\Gamma_j(x-t)-\theta_j(0)} \right] . \end{aligned} \tag{14}$$

It is easy to show that for this choice of basis functions equation (1) leads to

$$\frac{dV_n}{dx} + \frac{dV_n}{dt} = 0 . \tag{15}$$

Therefore for $m = 1$ the time evolution of the potential V_n is trivial because V_n is a function of $(x - t)$. The K-dV equation arises from choosing $m = 3$ in equation (13).

The first new non-trivial case is realised for $m = 5$. Equation (13) can be satisfied by the choice

$$\theta_j(t) = \theta_j(0) - 16\,\Gamma_j^5\, t \tag{16}$$

which leads to

$$\begin{aligned} B_{ij} &= (1/2)\frac{d^{i-1}}{dx^{i-1}}\left[e^{y_j} + (-)^{j+1} e^{-y_j} \right] , \\ y_j &= \Gamma_j\left(x - 16\,\Gamma_j^4\, t \right) + \theta_j(0) . \end{aligned} \tag{17}$$

A long and tedious algebra shows that the n-soliton solution defined by equations (1) and (16) satisfies

$$\left[\frac{d^5}{dx^5} - 10V_n\frac{d^3}{dx^3} - 20\frac{dV_n}{dx}\frac{d^2}{dx^2} + 30V_n^2\frac{d}{dx} + \frac{d}{dt} \right] V_n(x,t) = 0 . \tag{18}$$

The non-linear equation defined by equations (18) has been studied by Sawada and Kotera (1974) and Caudrey *et al* (1976).

The next possible choice of $m = 7$ leads to the time variation of the phase of the form

$$\theta_j(t) = \theta_j(0) - 64\,\Gamma_j^7\, t . \tag{19}$$

It can be shown that the n-soliton solution defined by equations (1) and (19) satisfies

$$\begin{aligned} &\frac{d^7V_n}{dx^7} - 14\,V_n\frac{d^5V_n}{dx^5} - 42\,\frac{dV_n}{dx}\frac{d^4V_n}{dx^4} - 70\,\frac{d^2V_n}{dx^2}\frac{d^3V_n}{dx^3} \\ &+ 70\,V_n^2\frac{d^3V_n}{dx^3} + 280\,V_n\frac{dV_n}{dx}\frac{d^2V_n}{dx^2} + 70\left(\frac{dV_n}{dx}\right)^3 - 140\,V_n^3\frac{dV_n}{dx} + \frac{dV_n}{dt} = 0 . \end{aligned} \tag{20}$$

A non-linear equation arises for all odd values of m ($m \geq 3$) in equation (13) for which the time dependence of the phase is given by

$$\theta_j(t) = \theta_j(0) - 2^{m-1}\,\Gamma_j^m\,t\,. \tag{21}$$

For this class of non-linear equations, the n-soliton solution is given by equations (1) and (23) or by equation (6) with y_j suitably modified to produce the appropriate time dependence of the phase.

Caudrey *et al* (1976) derived a hierarchy of K-dV equations using the inverse scattering method. They showed that the non-linear equation of order $(2m+1)$ with n solitons satisfies

$$\frac{dV_n}{dt} + \frac{dL_m}{dx} = 0 \tag{22}$$

where the operator L_m is to be found by solving the system of equations defined by

$$\frac{dL_j}{dx} = \left[\frac{d^3 L_{j-1}}{dx^3} - 4V_n\,\frac{dL_{j-1}}{dx} - 2\,\frac{dV_n}{dx}\,L_{j-1}\right] \tag{23}$$

for $j = 1, 2...m$ with the starting condition

$$L_0 = V_n\,. \tag{24}$$

It is easy to show that $j = 1$ leads to the K-dV equation and $j = 2$ and $j = 3$ lead to the non-linear equations given by equations (18) and (20) respectively. Further iterations of equation (23) lead to the higher members of the K-dV hierarchy. The n-soliton solution of this hierarchy given by Caudrey *et al* (1976) can be shown to be identical to the solutions discussed in this paper. An explicit solution of equation (23) for the operator L_k for arbitrary k remains to be discovered. Such an explicit solution for L_k would provide an explicit structure for non-linear equations of the K-dV type for arbitrary order.

In this paper we have constructed a hierarchy of higher order non-linear equations of the K-dV type generated from the set of equations first considered by Caudrey *et al* (1976). The non-linear equations which are members of this hierarchy possess the attractive feature that explicit expressions for n-soliton solutions may be given. Furthermore it can be shown that for all the members of this hierarchy the n and the $n+1$ soliton solutions are supersymmetric partners.

REFERENCES

A. A. Andrianov, N. V. Borisov and M. V. Ioffe, Phys. Lett. **105A** 19 (1984).

P. J. Caudrey, R. K. Dodd and J. D. Gibbon, Proc. R. Soc. Lond. **A351** 407 (1976).

B. Freedman and F. Cooper, Ann. Phys. NY **146** 262 (1983).

C. S. Gardner, J. M. Greene, M. D. Kruskal and R. M. Miura, Phys. Rev. Lett. **19** 1095 (1967).

I. Kay and H. E. Moses, J. Appl. Phys. **27** 1503 (1956).

W. Kwong and J. L. Rossner, Prog. Theor. Phys. Suppl. **86** 366 (1985).

K. Sawada and T. Kotera, Prog. Theor. Phys. **51** 1355 (1974).

A. C. Scott, F. Y. F. Chu and D. W.McLaughlin, Proc. I.E.E.E. **61** 1443 (1973).

C. V. Sukumar, J. Phys. A **18** 2917 (1985).

C. V. Sukumar, J. Phys. A **18** 2937 (1985).

C. V. Sukumar, J. Phys. A **19** 2297 (1986).

E. Witten, Nucl. Phys. B **188** 513 (1981).

Third International Colloquium on Differential Equations pp. 177-192 (1993)
D. Bainov and V. Covachev (Eds)

On integral and differential equations arising from probability distributions

Katsuo Takano

College of General Education,
Ibaraki University,
Mito, Ibaraki 310, Japan

1991 *Mathematics Subject Classification*. Primary 45E10, Secondary 34A25, 60E07.

1. Introduction

The probability density function of variance ratio distribution or the F - distribution with m and n degrees of freedom is as follows:

$$g_{m,n}(x) = \frac{\Gamma((m+n)/2)}{\Gamma(m/2)\Gamma(n/2)} \frac{x^{(m/2)-1}}{(1+x)^{(m+n)/2}}, \quad x > 0,$$

cf.[1. p.946. 26.6.1].

The variance ratio distribution is an important distribution in statistics. We consider the density function

$$(1.1) \qquad g_{\alpha,\beta}(x) = \frac{\Gamma(\alpha+\beta)}{\Gamma(\alpha)\Gamma(\beta)} \frac{x^{\alpha-1}}{(1+x)^{\alpha+\beta}}, \quad x > 0,$$

where α and β are any positive numbers.

It is known that the Laplace transform of the infinitely divisible probability distribution $G(x)$ on $[0,\infty)$ is represented in the form

$$(1.2) \qquad \int_0^\infty e^{-sx} dG(x) = \exp\{\int_0^\infty (e^{-sx} - 1)\frac{1}{x} dK(x)\},$$

(cf.[5]).

If a probability distribution $G(x)$ on $[0,\infty)$ is infinitely divisible and has a probability density $g(x)$, its density $g(x)$ satisfies an integral equation

(1.3) $$xg(x) = \int_0^x g(x-t)dK(t), \quad x > 0$$

with conditions that (c1) $K(x)$ is nondecreasing, (c2) $K(-0) = 0$, (c3) $\int_1^\infty 1/x\, dK(x) < \infty$ (cf.[5]).
Conversely, if a density function $g(x)$ satisfies the integral equation (1.3) with the conditions (c1), (c2) and (c3), the probability distribution function $G(x)$ on $[0,\infty)$ with the density function $g(x)$ is infinitely divisible. If $dK(x)$ is absolutely continuous and its density function is $k(x)$, then (1.2) and (1.3) are written as

(1.4) $$\int_0^\infty e^{-sx} dG(x) = \exp\{\int_0^\infty (e^{-sx} - 1)\frac{1}{x} k(x)dx\}.$$

(1.5) $$xg(x) = \int_0^x g(x-t)k(t)dt, \quad x > 0.$$

For the case $\alpha = 1$, the density function (1.1) is called the density function of the Pareto distribution and O. Thorin [10] first obtained the function $K(x)$ for the case where β is a non-integer. For the case where α is any positive number and β is a non-integer, M. J. Goovaerts, L. D'Hooge, N. De Pril [3] obtained the function $K(x)$.
The purpose of this paper is to get the function $K(x)$ for the case where α is any positive number and β is any positive integer.
Now we get the function $k(t)$ of the integral equation

(1.6) $$\frac{x^\alpha}{(1+x)^{\alpha+\beta}} = \int_0^x \frac{(x-t)^{\alpha-1}}{(1+x-t)^{\alpha+\beta}} k(t)dt, \quad x > 0$$

for any positive α, β.
As seen from the integral equation (1.5), we can ignore the normalizing constant number of the probability density function and hence in general we can consider the solution or the function $k(x)$ of the integral equation (1.5).
The author would like to emphasize that the idea obtained here can also apply to other probability distributions.

2. On the Laplace transform of the density function of the variance ratio distribution

We make use of the confluent hypergeometric function

(2.1) $$M(a,b,z) = 1 + \frac{a}{b}z + \frac{(a)_2}{(b)_2}\frac{z^2}{2!} + \cdots + \frac{(a)_n}{(b)_n}\frac{z^n}{n!} + \cdots ,$$

where $(a)_0 = 1,\ (b)_0 = 1,$

$(a)_n = a(a+1)(a+2)\cdots(a+n-1),$

$(b)_n = b(b+1)(b+2)\cdots(b+n-1),\ b \neq 0,-1,-2,\ldots .$

The series $M(a,b,z)$ converges uniformly in z in any compact set. From [4. (3.1.19), (1.3.1)] we have

$$\Gamma(a)U(a,b,s) = \int_0^\infty e^{-st}t^{a-1}(1+t)^{b-a-1}dt$$

and

(2.2) $$U(a,b,s) = \frac{\Gamma(1-b)}{\Gamma(1+a-b)}M(a,b,s) + \frac{\Gamma(b-1)}{\Gamma(a)}s^{1-b}M(1+a-b,2-b,s).$$

The gamma function $\Gamma(z)$ is defined on the whole complex plane except $z = 0,-1,-2,\ldots$. Let

(2.3) $$f_{\alpha,\beta}(s) = \int_0^\infty e^{-sx}g_{\alpha,\beta}(x)dx,\ Re\ s \geq 0.$$

Suppose $\beta = l + h,\ l = 0,1,2,\ldots,\ 0 < h < 1$. Then $\alpha = a > 0,\ \beta = 1 - b > 0$ and we have

$$f_{\alpha,\beta}(s) = \frac{\Gamma(\alpha+\beta)}{\Gamma(\beta)}U(\alpha,1-\beta,s)$$

(2.4) $$= M(\alpha,1-\beta,s) + \frac{\Gamma(\alpha+\beta)\Gamma(-\beta)}{\Gamma(\beta)\Gamma(\alpha)}\ s^\beta M(\alpha+\beta,1+\beta,s),$$

for $Re\ s \geq 0$. We take the principal branch of $\log s$ such as $s^\beta = e^{\beta\log s} > 0$ for $s > 0$.

If $s = -t + i\rho,\ t > 0,\ \rho > 0$, then

$$U^+(\alpha,1-\beta,-t) = \lim_{\rho\to+0}U(\alpha,1-\beta,-t+i\rho)$$

(2.5) $$= \frac{\Gamma(\beta)}{\Gamma(\alpha+\beta)}M(\alpha,1-\beta,-t) + \frac{\Gamma(-\beta)}{\Gamma(\alpha)}\ t^\beta e^{i\pi\beta}\ M(\alpha+\beta,1+\beta,-t).$$

If $s = -t - i\rho,\ t > 0,\ \rho > 0$, then

$$U^-(\alpha,1-\beta,-t) = \lim_{\rho\to+0}U(\alpha,1-\beta,-t-i\rho)$$

(2.6) $$= \frac{\Gamma(\beta)}{\Gamma(\alpha+\beta)}M(\alpha,1-\beta,-t) + \frac{\Gamma(-\beta)}{\Gamma(\alpha)}\ t^\beta e^{-i\pi\beta}\ M(\alpha+\beta,1+\beta,-t).$$

Let

$$Im\frac{\Gamma(\alpha+\beta)}{\Gamma(\beta)}U^{-}(\alpha,1-\beta,-t) = \frac{\Gamma(\alpha+\beta)\Gamma(-\beta)}{\Gamma(\beta)\Gamma(\alpha)}(-1)\sin\pi\beta\cdot t^{\beta}M(\alpha+\beta,1+\beta,-t)$$

$$= u(\alpha,\beta;t). \tag{2.7}$$

Then

$$Im\frac{\Gamma(\alpha+\beta)}{\Gamma(\beta)}U^{+}(\alpha,1-\beta,-t) = \frac{\Gamma(\alpha+\beta)\Gamma(-\beta)}{\Gamma(\beta)\Gamma(\alpha)}\sin\pi\beta\cdot t^{\beta}M(\alpha+\beta,1+\beta,-t)$$

$$= -u(\alpha,\beta;t).$$

Let

$$Re\frac{\Gamma(\alpha+\beta)}{\Gamma(\beta)}U^{+}(\alpha,1-\beta,-t) = M(\alpha,1-\beta,-t)$$

$$+\frac{\Gamma(\alpha+\beta)\Gamma(-\beta)}{\Gamma(\beta)\Gamma(\alpha)}\cos\pi\beta\cdot t^{\beta}M(\alpha+\beta,1+\beta,-t)$$

$$= v(\alpha,\beta;t). \tag{2.8}$$

We also have

$$Re\frac{\Gamma(\alpha+\beta)}{\Gamma(\beta)}U^{-}(\alpha,1-\beta,-t) = v(\alpha,\beta;t).$$

We have

$$\Gamma(-\beta) = \frac{(-1)^{l+1}}{h(1+h)\cdots(l+h)}\Gamma(1-h) \tag{2.9}$$

and so

$$u(\alpha,\beta;t) = \frac{\Gamma(\alpha+\beta)}{\Gamma(\beta)\Gamma(\alpha)}\frac{\Gamma(1-h)\sin\pi h}{h(1+h)\cdots(l+h)}t^{\beta}M(\alpha+\beta,1+\beta,-t), \tag{2.10}$$

$$v(\alpha,\beta;t) = M(\alpha,1-\beta,-t) - \frac{\Gamma(\alpha+\beta)}{\Gamma(\beta)\Gamma(\alpha)}\frac{\Gamma(1-h)\cos\pi h}{h(1+h)\cdots(l+h)}t^{\beta}M(\alpha+\beta,1+\beta,-t). \tag{2.11}$$

3. The function $k(x)$ for the case where α is a positive number and β is a non-integer

Suppose $\beta = l+h$, l is a non-negative integer, $0 < h < 1$.
If the density function (1.1) of the variance ratio distribution satisfies the integral equation (1.5), taking the Laplace transform of the integral equation (1.5), we have

$$-\frac{f'_{\alpha,\beta}(s)}{f_{\alpha,\beta}(s)} = \int_0^{\infty} e^{-sx}dK(x), \quad Re\, s > 0.$$

Calculating the inverse Laplace transform of $f'_{\alpha,\beta}(s)/f_{\alpha,\beta}(s)$, we can find the function $K(x)$ satisfying (c1), (c2) and (c3).

Let

$$c(\alpha,\beta) = \frac{\Gamma(\alpha+\beta)}{\Gamma(\beta)\Gamma(\alpha)}\frac{\Gamma(1-h)\sin\pi h}{h(1+h)\cdots(l+h)}.$$

Lemma 1. Let $p(\beta;t) = t^{1-\beta}e^t$ and $q(\alpha,\beta;t) = \alpha t^{-\beta}e^t$, $t > 0$. Then $u(\alpha,\beta;t)$ and $v(\alpha,\beta;t)$ are linearly independent solutions of the differential equation

(3.1) $$(p(\beta;t)y'(t))' + q(\alpha,\beta;t)y(t) = 0, \quad t > 0.$$

The Wronskian of $u(\alpha,\beta;t)$ and $v(\alpha,\beta;t)$ is as follows:

(3.2) $$p(\beta;t)(u'(\alpha,\beta;t)v(\alpha,\beta;t) - u(\alpha,\beta;t)v'(\alpha,\beta;t)) = c(\alpha,\beta)\beta.$$

Proof. The differential equation (3.1) is equivalent to the differential equation

(3.3) $$ty''(t) + (1-\beta+t)y'(t) + \alpha y(t) = 0, \quad t > 0.$$

We can make use of the fact that the functions $M(\alpha+\beta,1+\beta,t)$ and $M(\alpha,1-\beta,t)$ are solutions of Kummer's equations, $ty''(t) + (1+\beta-t)y'(t) - (\alpha+\beta)y(t) = 0$ and $ty''(t) + (1-\beta-t)y'(t) - \alpha y(t) = 0$, respectively.

With the Wronskian, $W(t) = u'(\alpha,\beta;t)v(\alpha,\beta;t) - u(\alpha,\beta;t)v'(\alpha,\beta;t)$, we have

$$\begin{aligned} W'(t) &= u''(\alpha,\beta;t)v(\alpha,\beta;t) - u(\alpha,\beta;t)v''(\alpha,\beta;t) \\ &= -\frac{1}{t}\{(1-\beta+t)u'(\alpha,\beta;t) + \alpha u(\alpha,\beta;t)\}v(\alpha,\beta;t) \\ &\quad + u(\alpha,\beta;t)\frac{1}{t}\{(1-\beta+t)v'(\alpha,\beta;t) + \alpha v(\alpha,\beta;t)\} \\ &= -\frac{1-\beta+t}{t}W(t). \end{aligned}$$

Hence we have

$W(t) = Ct^{\beta-1}e^{-t}$, $t > 0$, where C is a constant number,

and from the fact that

$$\lim_{t\to+0} t^{1-\beta}(u'(\alpha,\beta;t)v(\alpha,\beta;t) - u(\alpha,\beta;t)v'(\alpha,\beta;t)) = c(\alpha,\beta)\beta,$$

we obtain

$$W(t) = c(\alpha,\beta)\beta t^{\beta-1}e^{-t}, \quad t > 0.$$

q.e.d.

Lemma 2. If m is a positive integer such that $m < \alpha \le m+1$

then the confluent hypergeometric function $M(1-\alpha,1+\beta,t)$ has m simple positive zeros.

Proof. It follows from the expression

$$M(1-\alpha,1+\beta,t)$$
$$= 1+\frac{1-\alpha}{1+\beta}t+\frac{(1-\alpha)(2-\alpha)}{(1+\beta)(2+\beta)}\frac{t^2}{2!}+\cdots+\frac{(1-\alpha)(2-\alpha)\cdots(n-\alpha)}{(1+\beta)(2+\beta)\cdots(n+\beta)}\frac{t^n}{n!}+\cdots$$

that $M(1-\alpha,1+\beta,t)$ is larger than one for $t>0$ if $0<\alpha\le 1$. If $1<\alpha\le 2$, it holds that $M(1-\alpha,1+\beta,0)=1$ and

$$M'(1-\alpha,1+\beta,t)=\frac{1-\alpha}{1+\beta}M(2-\alpha,2+\beta,t)\le\frac{1-\alpha}{1+\beta},\quad t>0.$$

Hence $M(1-\alpha,1+\beta,t)$ has one positive zero.

For $2<\alpha\le 3$, we make use of the relation such as

$$b(1-b+t)M(a,b,t)+b(b-1)M(a-1,b-1,t)-atM(a+1,b+1,t)=0.$$

cf. [1. 13.4.7]. If we let $a=2-\alpha$, $b=2+\beta$, we have

$$(2+\beta)(1+\beta)M(1-\alpha,1+\beta,t)=b(b-1)M(a-1,b-1,t)$$
$$=-b(1-b+t)M(a,b,t)+atM(a+1,b+1,t)$$
$$=(2+\beta)(1+\beta-t)M(2-\alpha,2+\beta,t)-(\alpha-2)tM(3-\alpha,3+\beta,t).$$

If we denote the zero of $M(2-\alpha,2+\beta,t)$ by t_1, we have

$$(2+\beta)(1+\beta)M(1-\alpha,1+\beta,t_1)=-(\alpha-2)t_1M(3-\alpha,3+\beta,t_1)<0.$$

Furthermore we see that

$$M(1-\alpha,1+\beta,0)=1,$$
$$M'(1-\alpha,1+\beta,t)=\frac{1-\alpha}{1+\beta}M(2-\alpha,2+\beta,t),$$
$$M'(1-\alpha,1+\beta,0)=\frac{1-\alpha}{1+\beta}M(2-\alpha,2+\beta,0)\le\frac{1-\alpha}{1+\beta}<0,$$
$$M''(1-\alpha,1+\beta,t)=\frac{(1-\alpha)(2-\alpha)}{(1+\beta)(2+\beta)}M(3-\alpha,3+\beta,t)\ge\frac{(1-\alpha)(2-\alpha)}{(1+\beta)(2+\beta)}>0,\quad t>0.$$

Hence the function $M(1-\alpha,1+\beta,t)$ has two positive zeros.
If $m<\alpha\le m+1$, $m=3,4,\ldots$, repeating the above discussion inductively, we can show that the function $M(1-\alpha,1+\beta,t)$ has m positive zeros. Since $M(1-\alpha,1+\beta,t)$ is the solution of Kummer's equation, the zero of $M(1-\alpha,1+\beta,t)$ is simple. cf. [6. p.3]. q.e.d.

Theorem A. The function $k(x)$ is obtained by

$$(3.4)\quad k(x) = k(\alpha,\beta;x) = \int_0^{\infty} e^{-xt} \frac{c(\alpha,\beta)\beta}{\pi p(\beta;t)(u^2(\alpha,\beta;t) + v^2(\alpha,\beta;t))} dt, \quad x > 0$$

for the density function (1.1) with the positive number α and the positive non-integer β.

Proof. We suppose that $K(x)$ has a density function $k(\alpha,\beta;x)$ and let us calculate the inverse Laplace transform of $U'(\alpha,1-\beta,s)/U(\alpha,1-\beta,s)$. In what follows, we calculate the inverse Lapalce transform, that is,

$$k(\alpha,\beta;x) = \lim_{R\to\infty} \frac{1}{2\pi i}\int_{\delta-iR}^{\delta+iR} e^{xs}(-1)\frac{U'(\alpha,1-\beta,s)}{U(\alpha,1-\beta,s)} ds, \quad \delta > 0,\ x > 0.$$

Let us consider the contour integral along the closed curve in the figure. The function $U(\alpha,1-\beta,s)$ has no zeros outside the cut along the nonpositive real axis (cf. [11]) and $U'(\alpha,1-\beta,s)/U(\alpha,1-\beta,s)$ is an analytic function outside the cut and continuous on each side of the cut with exception of $s = 0$. Using the formula [4. (2.1.24), (4.1.5)],

$$\frac{d}{ds}U(a,b,s) = -aU(a+1,b+1,s),$$

$$U(a,b,s) = s^{-a}(1 + O(|s|^{-1})) \text{ for } |s| \text{ large and } |\arg s| < \frac{3\pi}{2},$$

we have

$$\left|\frac{U'(\alpha,1-\beta,s)}{U(\alpha,1-\beta,s)}\right| = \frac{\alpha|1 + O(|s|^{-1})|}{|s||1 + O(|s|^{-1})|}$$

for large $|s|$, $|\arg s| < 3\pi/2$. From this we have

$$\left|\int_{C_2} e^{xs}\frac{U'(\alpha,1-\beta,s)}{U(\alpha,1-\beta,s)} ds\right| \le \int_0^{\delta} |e^{x(\sigma+iR)}| \frac{\alpha}{|\sigma+iR|} |1 + O(|\sigma + iR|^{-1})| d\sigma \to 0$$

and

$$\left|\int_{C_3} e^{xs}\frac{U'(\alpha,1-\beta,s)}{U(\alpha,1-\beta,s)} ds\right| \le \alpha \int_{\pi/2}^{\pi} |e^{xRe^{i\theta}}| |1 + O(|Re^{i\theta}|^{-1})| d\theta \to 0$$

as $R\to\infty$.

Since $U(\alpha,1-\beta,re^{i\theta}) \to \frac{\Gamma(\alpha)}{\Gamma(\alpha+\beta)}$, $U'(\alpha,1-\beta,\ re^{i\theta})re^{i\theta} \to 0$ as $r\to 0$

$-\pi < \theta < \pi$, we have

$$\left|\int_{C_5} e^{xs}\frac{U'(\alpha,1-\beta,s)}{U(\alpha,1-\beta,s)}ds\right|\to 0$$

as $r\to 0$. The same facts hold for the integrals along C_7, C_8.

Let us consider along C_4, C_6. From the fact

that $u(\alpha,\beta;t)$ and $v(\alpha,\beta;t)$ do not vanish simultaneously and

$$\lim_{\rho\to+0} U'(\alpha,1-\beta,-t+i\rho) = -\frac{d}{dt}\lim_{\rho\to+0} U(\alpha,1-\beta,-t+i\rho),$$

$$\lim_{\rho\to+0} U'(\alpha,1-\beta,-t-i\rho) = -\frac{d}{dt}\lim_{\rho\to+0} U(\alpha,1-\beta,-t-i\rho),$$

we see that

(3.5)

$$\lim_{R\to\infty,\, r\to 0}\left\{\int_{C_4} e^{xs}\frac{U'(\alpha,1-\beta,s)}{U(\alpha,1-\beta,s)}ds + \int_{C_6} e^{xs}\frac{U'(\alpha,1-\beta,s)}{U(\alpha,1-\beta,s)}ds\right\}$$

$$= \lim_{R\to\infty,\, r\to 0}\left\{\int_{-R}^{-r} e^{x\sigma}\frac{U'^{+}(\alpha,1-\beta,\sigma)}{U^{+}(\alpha,1-\beta,\sigma)}d\sigma - \int_{-R}^{-r} e^{x\sigma}\frac{U'^{-}(\alpha,1-\beta,\sigma)}{U^{-}(\alpha,1-\beta,\sigma)}d\sigma\right\}$$

$$= \lim_{R\to\infty,\, r\to 0}\int_{r}^{R} e^{-xt}$$

$$\frac{U'^{+}(\alpha,1-\beta,-t)U^{-}(\alpha,1-\beta,-t) - U^{+}(\alpha,1-\beta,-t)U'^{-}(\alpha,1-\beta,-t)}{U^{+}(\alpha,1-\beta,-t)U^{-}(\alpha,1-\beta,-t)}dt$$

$$= 2i\lim_{R\to\infty,\, r\to 0}\int_{r}^{R} e^{-xt}\frac{u'(\alpha,\beta;t)v(\alpha,\beta;t) - u(\alpha,\beta;t)v'(\alpha,\beta;t)}{u^2(\alpha,\beta;t) + v^2(\alpha,\beta;t)}dt$$

$$= 2i\int_{0}^{\infty} e^{-xt}\frac{c(\alpha,\beta)\beta}{p(\beta;t)(u^2(\alpha,\beta;t) + v^2(\alpha,\beta;t))}dt.$$

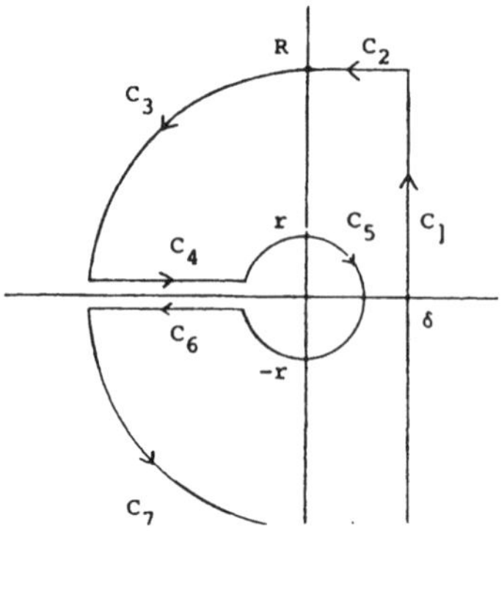

Let us put $q(\alpha,\beta;t) = v(\alpha,\beta;t)/u(\alpha,\beta;t)$.
By Kummer's transformation, we have
$u(\alpha,\beta;t) = c(\alpha,\beta)t^{\beta}M(\alpha+\beta,1+\beta,-t) = c(\alpha,\beta)t^{\beta}e^{-t}M(1-\alpha,1+\beta,t)$.
cf. [4. (1.4.1)]. If m is a positive integer and $m < \alpha \leq m+1$, the confluent hypergeometric function $M(1-\alpha,1+\beta,t)$ has m simple real positive zeros.
Since the functions $u(\alpha,\beta;t)$ and $v(\alpha,\beta;t)$ are the solutions of the differential eqution (3.1), the Sturm separation theorem tells that zeros of $u(\alpha,\beta;t)$ and the zeros of $v(\alpha,\beta;t)$ separate each other except $t = 0$. cf. [6]. It holds that $q(\alpha,\beta;t)\to+\infty$ as $t\to+0$ and

$$q'(\alpha,\beta;t) = -\frac{c(\alpha,\beta)\beta}{p(\beta;t)u^2(\alpha,\beta;t)} < 0$$

for $t > 0$ except for the zeros of the solution $u(\alpha,\beta;t)$.
If we take $\tan^{-1}u(\alpha,\beta;+0)/v(\alpha,\beta;+0) = 0$, there exists a positive number $m(\alpha,\beta)$ less than or equal to $(m+1)\pi$ such that

$$\text{(3.6)} \qquad \tan^{-1}\frac{u(\alpha,\beta;t)}{v(\alpha,\beta;t)}\to m(\alpha,\beta), \quad (t\to+\infty).$$

Thus, from (3.5) we have

$$\text{(3.7)} \quad \pi k(\alpha,\beta;x) = \int_0^{\infty} e^{-xt}\frac{c(\alpha,\beta)\beta}{p(\beta;t)(u^2(\alpha,\beta;t)+v^2(\alpha,\beta;t))}dt$$

$$= \lim_{R\to\infty,\ r\to+0}\{e^{-xR}\tan^{-1}\frac{u(\alpha,\beta;R)}{v(\alpha,\beta;R)} - e^{-xr}\tan^{-1}\frac{u(\alpha,\beta;r)}{v(\alpha,\beta;r)} + x\int_r^R e^{-xt}\tan^{-1}\frac{u(\alpha,\beta;t)}{v(\alpha,\beta;t)}dt\}$$

$$= x\int_0^{\infty} e^{-xt}\tan^{-1}\frac{u(\alpha,\beta;t)}{v(\alpha,\beta;t)}dt, \quad (x > 0),$$

and so $k(\alpha,\beta;x)$ is a finite valued function.
By the fact that as $t\to+0$,

$$\tan^{-1}\frac{u(\alpha,\beta;t)}{v(\alpha,\beta;t)}\sim c(\alpha,\beta)t^{\beta}$$

we see that

$$\int_1^{\infty}\frac{1}{x}k(\alpha,\beta;x)dx = \frac{1}{\pi}\int_0^{\infty}\frac{1}{t}e^{-t}\tan^{-1}\frac{u(\alpha,\beta;t)}{v(\alpha,\beta;t)}dt < \infty.$$

q.e.d.

4. The limiting function of $u(\alpha,\beta;t)$ and $v(\alpha,\beta;t)$ as β approaches a positive integer l.

Suppose that α is any positive number and $\beta = l+h$, l is a positive integer, $0 < h < 1$.
As h approaches +0, then $p(\beta;t) = t^{1-\beta}e^t \to p(l;t) = t^{1-l}e^t$ and $q(\alpha,\beta;t) = \alpha t^{-\beta}e^t \to q(\alpha,l;t) = \alpha t^{-l}e^t$, $t > 0$.
As h approaches +0 then
$\frac{1}{\pi}c(\alpha,\beta)$ approaches $c(l) = \Gamma(\alpha+l)/(\Gamma(l)\Gamma(\alpha)l!)$.
As h approaches +0, the function $u(\alpha,\beta;t)$ converges to $u(\alpha,l;t) = c(l)\pi t^l M(\alpha+l,1+l,-t)$ uniformly in t in any compact set.
We can rewrite (2.11) as follows:

$$(4.1)\quad v(\alpha,\beta;t) = 1+\frac{\alpha}{\beta-1}t+\frac{\alpha(\alpha+1)}{(\beta-1)(\beta-2)}\frac{t^2}{2!}+\cdots+\frac{\alpha(\alpha+1)\cdots(\alpha+l-2)}{(\beta-1)(\beta-2)\cdots(\beta-l+1)}\frac{t^{l-1}}{(l-1)!}$$
$$+\frac{\alpha(\alpha+1)\cdots(\alpha+l-1)}{(\beta-1)(\beta-2)\cdots(1+h)h}\frac{t^l}{l!}[(1+\frac{\alpha+l}{l+1}(-t)\frac{1}{(1-h)}+\frac{(\alpha+l)(\alpha+l+1)}{(l+1)(l+2)}\frac{(-t)^2}{2!}\frac{1}{(1-h)(1-h/2)}$$
$$+\frac{(\alpha+l)(\alpha+l+1)(\alpha+l+2)}{(l+1)(l+2)(l+3)}\frac{(-t)^3}{3!}\frac{1}{(1-h)(1-h/2)(1-h/3)}+\cdots)-M(\alpha+l,1+l,-t)]$$
$$+\frac{\alpha(\alpha+1)\cdots(\alpha+l-1)}{(\beta-1)(\beta-2)\cdots(1+h)h}\frac{t^l}{l!}M(\alpha+l,1+l,-t)$$
$$-\frac{\Gamma(\alpha+\beta)}{\Gamma(\beta)\Gamma(\alpha)}\frac{\Gamma(1-h)\cos\pi h}{h(1+h)\cdots(l+h)}t^\beta M(\alpha+\beta,1+\beta,-t).$$

Lemma 3. As h approaches +0, then the function $v(\alpha,\beta;t)$ converges to

$$(4.2)\quad v(\alpha,l;t) = 1+\frac{\alpha}{l-1}t+\frac{\alpha(\alpha+1)}{(l-1)(l-2)}\frac{t^2}{2!}+\cdots+\frac{\alpha(\alpha+1)\cdots(\alpha+l-2)}{(l-1)(l-2)\cdots2\cdot1}\frac{t^{l-1}}{(l-1)!}$$
$$+c(l)t^l\Sigma_{r=1}^{\infty}\frac{(\alpha+l)_r}{(1+l)_r}\frac{(-t)^r}{r!}\{\Sigma_{j=0}^{r-1}(\frac{1}{1+j}+\frac{1}{1+l+j}-\frac{1}{\alpha+l+j})\}$$
$$-c(l)t^l M(\alpha+l,1+l,-t)(\log t+A(l))$$

for every positive t,

where $A(l) = \frac{\Gamma'(\alpha+l)}{\Gamma(\alpha+l)} - \frac{\Gamma'(l)}{\Gamma(l)} - \Gamma'(1) - \frac{1}{l}$.

If $\alpha = 1$ we have

$M(\alpha + l, 1 + l, -t) = e^{-t}$, $A(l) = -\Gamma'(1)$.

Proof. Suppose $\alpha \neq 1$. It holds that

$$(4.3)\quad \frac{\alpha(\alpha+1)\cdots(\alpha+l-1)}{(\beta-1)(\beta-2)\cdots(1+h)h}\frac{t^l}{l!}[(1 + \frac{\alpha+l}{l+1}(-t)\frac{1}{(1-h)} + \frac{(\alpha+l)(\alpha+l+1)}{(l+1)(l+2)}\frac{(-t)^2}{2!}\frac{1}{(1-h)(1-h/2)}$$
$$+ \frac{(\alpha+l)(\alpha+l+1)(\alpha+l+2)}{(l+1)(l+2)(l+3)}\frac{(-t)^3}{3!}\frac{1}{(1-h)(1-h/2)(1-h/3)} + \cdots) - M(\alpha + l, 1 + l, -t)]$$
$$\to c(l)t^l\Sigma_{r=1}^{\infty}\frac{(\alpha+l)_r}{(1+l)_r}\frac{(-t)^r}{r!}\Sigma_{j=0}^{r-1}\frac{1}{1+j}$$

as $h \to +0$ uniformly in t in any compact set. It holds that

$$(4.4)\quad \frac{\alpha(\alpha+1)\cdots(\alpha+l-1)}{(\beta-1)(\beta-2)\cdots(1+h)h}\frac{t^l}{l!}M(\alpha + l, 1 + l, -t)$$
$$-\frac{\Gamma(\alpha+\beta)}{\Gamma(\beta)\Gamma(\alpha)}\frac{\Gamma(1-h)\cos\pi h}{h(1+h)\cdots(l+h)}t^{\beta}M(\alpha + \beta, 1 + \beta, -t).$$
$$= \{\frac{\Gamma(\alpha+l)}{\Gamma(l)\Gamma(\alpha)} - \frac{\Gamma(\alpha+\beta)}{\Gamma(\beta)\Gamma(\alpha)}\}\frac{\Gamma(1-h)\cos\pi h}{h(1+h)\cdots(l+h)}t^{\beta}M(\alpha + \beta, 1 + \beta, -t)$$
$$+ \frac{\Gamma(\alpha+l)}{\Gamma(l)\Gamma(\alpha)}\frac{(\Gamma(1)-\Gamma(1-h))\cos\pi h}{h(1+h)\cdots(l+h)}t^{\beta}M(\alpha + \beta, 1 + \beta, -t)$$
$$+ \frac{\Gamma(\alpha+l)}{\Gamma(l)\Gamma(\alpha)}\frac{1-\cos\pi h}{h(1+h)\cdots(l+h)}t^{\beta}M(\alpha + \beta, 1 + \beta, -t)$$
$$+ \frac{\Gamma(\alpha+l)}{\Gamma(l)\Gamma(\alpha)}\frac{t^l}{h(1+h)\cdots(l+h)}(1 - t^h)M(\alpha + \beta, 1 + \beta, -t)$$
$$+ \frac{\Gamma(\alpha+l)}{\Gamma(l)\Gamma(\alpha)}\frac{t^l}{h(1+h)\cdots(l+h)}\{M(\alpha + l, 1 + \beta, -t) - M(\alpha + l + h, 1 + \beta, -t)\}$$
$$+ \frac{\Gamma(\alpha+l)}{\Gamma(l)\Gamma(\alpha)}\frac{t^l}{h(1+h)\cdots(l+h)}\{M(\alpha + l, 1 + l, -t) - M(\alpha + l, 1 + l + h, -t)\}$$
$$+ \frac{\Gamma(\alpha+l)}{\Gamma(l)\Gamma(\alpha)}(\frac{1}{l} - \frac{1}{l+h})\frac{t^l}{h(1+h)\cdots(l-1+h)}M(\alpha + l, 1 + l, -t).$$

We have

$$(4.5)\quad \{\frac{\Gamma(\alpha+l)}{\Gamma(l)\Gamma(\alpha)} - \frac{\Gamma(\alpha+\beta)}{\Gamma(\beta)\Gamma(\alpha)}\}\frac{\Gamma(1-h)\cos\pi h}{h(1+h)\cdots(l+h)}t^{\beta}M(\alpha + \beta, 1 + \beta, -t)$$
$$\to -\{\Gamma'(\alpha + l)\Gamma(l) - \Gamma(\alpha + l)\Gamma'(l)\}/\{\Gamma(\alpha)\Gamma(l)^2\}\frac{t^l}{l!}M(\alpha + l, 1 + l, -t)$$

$$= -c(l)\{\frac{\Gamma'(\alpha+l)}{\Gamma(\alpha+l)} - \frac{\Gamma'(l)}{\Gamma(l)}\}t^l M(\alpha + l, 1 + l, -t)$$

as $h \to +0$. We have

$$(4.6)\quad \frac{\Gamma(\alpha+l)}{\Gamma(l)\Gamma(\alpha)} \frac{(\Gamma(1)-\Gamma(1-h))\cos \pi h}{h(1+h)\cdots(l+h)} t^{\beta} M(\alpha + \beta, 1 + \beta, -t)$$
$$\to c(l)\Gamma'(1)t^l M(\alpha + l, 1 + l, -t)$$

as $h \to +0$. We have

$$(4.7)\quad \frac{\Gamma(\alpha+l)}{\Gamma(l)\Gamma(\alpha)} \frac{1-\cos \pi h}{h(1+h)\cdots(l+h)} t^{\beta} M(\alpha + \beta, 1 + \beta, -t) \to 0$$

as $h \to +0$. We have

$$(4.8)\quad \frac{\Gamma(\alpha+l)}{\Gamma(l)\Gamma(\alpha)} \frac{t^l}{h(1+h)\cdots(l+h)} (1 - t^h) M(\alpha + \beta, 1 + \beta, -t)$$
$$\to -c(l)t^l \log t \cdot M(\alpha + l, 1 + l, -t)$$

as $h \to +0$ for every positive t. We have

$$(4.9)\quad \frac{\Gamma(\alpha+l)}{\Gamma(l)\Gamma(\alpha)} \frac{t^l}{h(1+h)\cdots(l+h)} \{M(\alpha + l, 1 + \beta, -t) - M(\alpha + l + h, 1 + \beta, -t)\}$$
$$\to -c(l)t^l \Sigma_{r=1}^{\infty} \frac{(\alpha+l)_r}{(1+l)_r} \frac{(-t)^r}{r!} \Sigma_{j=0}^{r-1} \frac{1}{\alpha+l+j}$$

as $h \to +0$ uniformly in t in any compact set. We have

$$(4.10)\quad \frac{\Gamma(\alpha+l)}{\Gamma(l)\Gamma(\alpha)} \frac{t^l}{h(1+h)\cdots(l+h)} \{M(\alpha + l, 1 + l, -t) - M(\alpha + l, 1 + l + h, -t)\}$$
$$\to c(l)t^l \Sigma_{r=1}^{\infty} \frac{(\alpha+l)_r}{(1+l)_r} \frac{(-t)^r}{r!} \Sigma_{j=0}^{r-1} \frac{1}{l+j+1}$$

as $h \to +0$ uniformly in t in any compact set. We have

$$(4.11)\quad \frac{\Gamma(\alpha+l)}{\Gamma(l)\Gamma(\alpha)} (\frac{1}{l} - \frac{1}{l+h}) \frac{t^l}{h(1+h)\cdots(l-1+h)} M(\alpha + l, 1 + l, -t)$$
$$\to \frac{c(l)}{l} t^l M(\alpha + l, 1 + l, -t)$$

as $h \to +0$.

Combining the above (4.3)-(4.11), we obtain (4.2).

If $\alpha = 1$ then

$M(\alpha + l, 1 + l, -t) = M(\alpha + \beta, 1 + \beta, -t) = e^{-t}$,

and so we obtain the assertion. q.e.d.

5. The function $k(x)$ for the case where α is a positive number and β is a positive integer l

Lemma 4. The functions $u(\alpha,l;t)$ and $v(\alpha,l;t)$ are solutions of the differential equation

(5.1) $$\{p(l;t)y'(t)\}' + q(\alpha,l;t)y(t) = 0, \quad t > 0,$$

with the Wronskian

(5.2) $$p(l;t)(u'(\alpha,l;t)v(\alpha,l;t) - u(\alpha,l;t)v'(\alpha,l;t)) = c(l)l\pi.$$

Proof. The differential equation (5.1) is equivalent to the differential equation

(5.3) $$ty''(t) + (1 - l + t)y'(t) + \alpha y(t) = 0, \quad t > 0.$$

By usual calculations of differentiations we can show that $u(\alpha,l;t)$ and $v(\alpha,l;t)$ are the solutions of the equation (5.3). In the same way as the proof of Lemma 1 we can obtain (5.2). q.e.d.

Theorem B. The function $k(x)$ is given by

(5.4) $$k(x) = k(\alpha,l;x) = \int_0^\infty e^{-xt} \frac{c(l)l}{p(l;t)(u^2(\alpha,l;t)+v^2(\alpha,l;t))} dt, \quad x > 0$$

for the density function (1.1) with the positive number α and the positive integer l.

Proof. Suppose that $\beta = l + h$, l is a positive integer, $0 < h < 1$. Let us show that $k(\alpha,\beta;x) \to k(\alpha,l;x)$ as $h \to +0$.
Let us put $q(\alpha,l;t) = v(\alpha,l;t)/u(\alpha,l;t)$.
By Kummer's transformation, we have
$u(\alpha,l;t) = c(l)\pi t^l M(\alpha + l, 1 + l, -t) = c(l)\pi t^l e^{-t} M(1 - \alpha, 1 + l, t)$.
cf. [4. (1.4.1)]. If m is a positive integer and $m < \alpha \le m + 1$, the confluent hypergeometric function $M(1 - \alpha, 1 + l, t)$ has m simple real positive zeros. Since the functions $u(\alpha,l;t)$ and $v(\alpha,l;t)$ are the solutions of the differential eqution (5.1), the Sturm separation theorem tells that the zeros of $u(\alpha,l;t)$ and the zeros of

$v(\alpha,l;t)$ separate each other except $t = 0$ (cf. [6]).
It holds that $q(\alpha,l;t) \to +\infty$ as $t \to +0$ and

$$q'(\alpha,l;t) = -\frac{c(l)l\pi}{p(l;t)u^2(\alpha,l;t)} < 0,$$

for $t > 0$ except for the zeros of the solution $u(\alpha,l;t)$.
If we take $\tan^{-1} u(\alpha,l;+0)/v(\alpha,l;+0) = 0$, there is a positive number $m(\alpha,l)$ such that

$$\tan^{-1}\frac{u(\alpha,l;t)}{v(\alpha,l;t)} \to m(\alpha,l) \le (m+1)\pi, \quad (t \to +\infty).$$

By integration by parts we have

$$k(\alpha,l;x) = \int_0^\infty e^{-xt}\frac{c(l)l}{p(l;t)(u^2(\alpha,l;t)+v^2(\alpha,l;t))}dt$$

$$= \lim_{R\to\infty,\ r\to+0}\frac{1}{\pi}\{e^{-xR}\tan^{-1}\frac{u(\alpha,l;R)}{v(\alpha,l;R)} - e^{-xr}\tan^{-1}\frac{u(\alpha,l;r)}{v(\alpha,l;r)} + x\int_r^R e^{-xt}\tan^{-1}\frac{u(\alpha,l;t)}{v(\alpha,l;t)}dt\}$$

$$= \frac{x}{\pi}\int_0^\infty e^{-xt}\tan^{-1}\frac{u(\alpha,l;t)}{v(\alpha,l;t)}dt, \ (x > 0),$$

and so $k(\alpha,l;x)$ is a finite valued function.
By the fact that as $h \to +0$

$$\tan^{-1}\frac{u(\alpha,l+h;t)}{v(\alpha,l+h;t)} \to \tan^{-1}\frac{u(\alpha,l;t)}{v(\alpha,l;t)}$$

for any t such that $v(\alpha,l,t) \ne 0$ and by the dominated convergence theorem we obtain

$k(\alpha,\beta;x) \to k(\alpha,l;x)$, $(x > 0)$

as $h \to +0$. The function $k(\alpha,l;x)$ satisfies the condition (c3).
As $h \to +0$,

$$f_{\alpha,\beta}(s) \to f_{\alpha,l}(s), \ Re\, s \ge 0,$$

$$f'_{\alpha,\beta}(s) \to f'_{\alpha,l}(s), \ Re\, s > 0,$$

hold and we have

$$-f'_{\alpha,\beta}(s) = f_{\alpha,\beta}(s)\int_0^\infty e^{-sx}k(\alpha,\beta;x)dx$$

$$\to -f'_{\alpha,l}(s) = f_{\alpha,l}(s)\int_0^\infty e^{-sx}k(\alpha,l;x)dx$$

for $Re\, s > 0$. By the inversion formula of the Laplace transform we obtain

$$xg_{\alpha,l}(x) = \int_0^x g_{\alpha,l}(x-t)k(\alpha,l;t)dt.$$

q.e.d.

Acknowledgement. The author would like to thank Prof. I. P. Stavroulakis for his suggestion concerning the Sturm-Picone theorem.

References

[1] M. Abramovitz and I. A. Stegun, Handbook of Mathematical Functions, New York, Dover, 1970.

[2] W. Feller, An introduction to probability theory and its applications, Vol. II. John Wiley, New York, 1971.

[3] M. J. Goovaerts, L. D'Hooge and N. de. Pril, On the infinite divisibility of the ratio of two gamma-distributed variables, Stochastic Processes and their Applications 7 1978, 292--297.

[4] L. J. Slater, Confluent hypergeometric functions, Cambridge University Press, London, 1960.

[5] F. W. Steutel, Preservation of infinite divisibility under mixing and related topics, Math. Centre Tracts 33. Math. Centre, Amsterdam, 1970.

[6] C. A. Swanson, Comparison and oscillation theory of linear differentail equations, Academic Press, New York, 1968.

[7] K. Sato and M. Yamazato, On distribution functions of class L, Z. Wahrscheinlichkeitstheorie verw. Gebiete 43 1978, 273--308.

[8] K. Takano, On global properties of solutions of a certain delay differential equation, J. London Math. Soc. 28 1983, 519--530.

[9] ------, On relations between the Pareto distribution and a differential equation, Bull. Fac. Sci. Ibaraki Univ., Math. 24 1992, 15--30.

[10] O. Thorin, On the infinite divisibility of the Pareto distribution, Scand. Acturial. J. 1977, 31--40.

[11] F. G. Tricomi, Über die Abzählung der Nullstellen der konfluenten hypergeometrischen Funktionen, Math. Zeit. 52 1950, 669--675.

[12] G. N. Watson, A Treatise on the Theory of Bessel Functions, Second Edition, Cambridge University Press, 1966.

Third International Colloquium on Differential Equations pp. 193-203 (1993)
D. Bainov and V. Covachev (Eds)

SINGULARITIES FOR MONGE-AMPERE EQUATIONS

Mikio TSUJI

Department of Mathematics, Kyoto Sangyo University
Kamigamo, Kita-ku, Kyoto 603, Japan

Abstract. It is well known that Monge-Ampère equations have locally smooth solutions in general. If we extend the smooth solutions, the singularities may appear. Our aim is to consider the structure of the singularities of solutions which appear in the process of extension.

§1. Introduction. In this talk we consider the Cauchy problem for Monge-Ampère equations. For un known function $z=z(x,y)$ defined for $(x,y) \in R^2$, we denote $p=\partial z/\partial x$, $q=\partial z/\partial y$, $r=\partial^2 z/\partial x^2$, $s=\partial^2 z/\partial x \partial y$ and $t=\partial^2 z/\partial y^2$. Monge-Ampère equations are written as

$$(1) \quad F(x,y,z,p,q,r,s,t) = Ar + Bs + Ct + D(rt-s^2) - E = 0$$

where A, B, C, D and E are smooth functions of (x,y,z,p,q). If the equation (1) is not degenerate, it is well known that the Cauchy problem for (1) has locally smooth solutions (for example, [1], [2], [3], [4], [5], [6], [7]). Our interest is on the global behavior of solutions. But we can not expect that the Cauchy problem admits a smooth solutions in the large. This suggests that, if we extend the smooth solutions, the singularities may appear. The aim of this talk is to construct the singularities of solutions in certain hyperbolic case. For our aim, we have to represent the solutions explicitely. To do so, we apply the characteristic method developed principally by D. Darboux and E. Goursat ([2], [3], [4]). As it seems to us that the method is not familiar today, let us start to recall it briefly. In this note we will sketch our discussions without proofs. The detailed paper will be published elsewhere soon.

§2. Characteristic strips and intermediate integrals. Let

$$C:\ (x,y,z,p,q) = (x(\alpha),\ y(\alpha),\ z(\alpha),\ p(\alpha),\ q(\alpha)) , \quad \alpha \in R^1 ,$$

be a smooth curve in R^5, and suppose that it satisfies a strip condition

$$(2) \qquad \frac{dz}{d\alpha}(\alpha) = P(\alpha)\frac{dx}{d\alpha}(\alpha) + q(\alpha)\frac{dy}{d\alpha}(\alpha) .$$

As a "characteristic strip" for (1) means that one can not determine the values of the second derivatives of solution along the curve C, we have the following

Definition 1. A curve C concerning (x,y,z,p,q) is called a "characteristic strip" of (1) if it satisfies (2) and

$$(3) \qquad \det \begin{pmatrix} \frac{\partial F}{\partial r} & \frac{\partial F}{\partial s} & \frac{\partial F}{\partial t} \\ \dot{x} & \dot{y} & 0 \\ 0 & \dot{x} & \dot{y} \end{pmatrix} = 0$$

where $\dot{x} = dx/d\alpha$ and $\dot{y} = dy/d\alpha$.

The equation (3) is written as

$$(3)' \qquad \frac{\partial F}{\partial t}\dot{x}^2 - \frac{\partial F}{\partial s}\dot{x}\dot{y} + \frac{\partial F}{\partial r}\dot{y}^2 = 0 .$$

Denote the discriminant of (3)' by Δ, then

$$\Delta = F_s^2 - 4F_rF_t = B^2 - 4(AC+DE) .$$

If $\Delta < 0$, the the equation (1) is called to be elliptic. If $\Delta > 0$, the equation (1) is hyperbolic. In this talk, we will treat the equations of hyperbolic type. More precisely, we assume that $\Delta \geqq 0$, and that the equations have two independent first intergrals. This class of equations has been well studied especially by E. Goursat [3]. Let us eliminate the unknown parameters (r,s,t) from the equation (3). Denote by λ_1 and λ_2

the solutions of second order polynomial: $\lambda^2 + B\lambda + (AC+DE) = 0$. Then the characteristic strip satisfies the following equations:

$$(I)\quad \begin{cases} dz - pdx - qdy = 0 \\ Ddp + Cdx + \lambda_1 dy = 0 \\ Ddq + \lambda_2 dx + Ady = 0 \end{cases}, \quad \text{or} \quad (II)\quad \begin{cases} dz - pdx - qdy = 0 \\ Ddp + Cdx + \lambda_2 dy = 0 \\ Ddq + \lambda_1 dx + Ady = 0 \end{cases}.$$

Definition 2. A function $V=V(x,y,z,p,q)$ is called the "first integral" of (I) (or (II)) if it is constant on any solution of (I) (or (II) respectively).

Proposition 1. (Darboux [2], Goursat [3], [4]) The necessary and sufficient condition so that $V=V(x,y,z,p,q)$ be the first integral of (I) is that V is a solution of the following system of linear first order equations:

$$(4)\quad \begin{cases} L_1 V \underset{\text{def}}{=} D\dfrac{\partial V}{\partial x} + pD\dfrac{\partial V}{\partial z} - C\dfrac{\partial V}{\partial p} - \lambda_1 \dfrac{\partial V}{\partial q} = 0, \\ L_2 V \underset{\text{def}}{=} D\dfrac{\partial V}{\partial y} + qD\dfrac{\partial V}{\partial z} - \lambda_2 \dfrac{\partial V}{\partial p} - A\dfrac{\partial V}{\partial q} = 0. \end{cases}$$

Definition 3. Let u and v be the first integrals of (I) which are independent each other, and ϕ be any function of two variables whose gradient is not zero. Then $\phi(u,v)$ is called an "intermediate integral" of (1).

Using the above classical notions, we try to solve the Cauchy problem for (1) explicitely. The initial strip is given by

$$(5)\qquad S_0: (x,y,z,p,q) = (x_0(\alpha), y_0(\alpha), z_0(\alpha), p_0(\alpha), q_0(\alpha))$$

where $\dot{z}_0(\alpha)=p_0(\alpha)\,\dot{x}_0(\alpha) + q_0(\alpha)\,\dot{y}_0(\alpha)$. Let $z=z(x,y)$ be a solution of (1) defined in a domain U. If the initial strip S_0 is contained in a set $\{(x,y,z(x,y),\partial z/\partial x(x,y),\partial z/\partial y(x,y));\ (x,y) \in U\}$, we say that $z=z(x,y)$ satisfies the initial condition (5). The Cauchy problem (1)-(5) is to find an open domain U where there exists a function $z=z(x,y)$ satisfying the equation $F=0$ in U and the initial condition (5).

Now suppose the hypotheses:

(H1) The system of the equations (I) (or (II)) has two independent first integrals, and denote them by u and v.

(H2) The initial strip S_0 is not characteristic.

Remark. Let us write $L_3=L_1L_2-L_2L_1 \underset{\text{def}}{=} [L_1,L_2]$. The necessary and sufficient condition for (H1) is that any bracket $[L_i,L_j]$ $(1\leq i,j\leq 3)$ is a linear combination of L_1, L_2 and L_3.

Denote $u_0(\alpha)=u|_{S_0}$ and $v_0=v|_{S_0}$, and $T=\{(u,v);\ u=u_0(\alpha),\ v=v_0(\alpha)\}$. If $(u_0'(\alpha),v_0'(\alpha))\neq(0,0)$, it is easy to get locally a function $\phi(u,v)$ satisfying $\phi(u_0(\alpha),v_0(\alpha))=0$ and $(\text{grad}\ \phi)(u_0(\alpha),v_0(\alpha))\neq(0,0)$. But we would like to develop the global theory. Therefore $\phi(u,v)$ should be defined in the large with the above two properties. To prove the existence of such a function, we assume the following condition:

(H3) The curve T is simple and $(u_0'(\alpha),v_0'(\alpha))\neq(0,0)$. The sets $\{\alpha;u_0'(\alpha)=0\}$ and $\{\alpha;v_0'(\alpha)=0\}$ have not any point of accumulation.

Here we define $g(x,y,z,p,q) \underset{\text{def}}{=} \phi(u(x,y,z,p,q),v(x,y,z,p,q))$, and consider the Cauchy problem for $g(x,y,z,p,q)$ as follows:

$$(6)\quad \begin{cases} g(x,y,z,\partial z/\partial x,\partial z/\partial y) = 0 , \\ (z,\partial z/\partial x,\partial z/\partial y)\big|_{(x,y)=(x_0(\alpha),y_0(\alpha))} = (z_0(\alpha),p_0(\alpha),q_0(\alpha)). \end{cases}$$

Then G. Darboux [2] and E. Goursat [3,4] have proved the following

Theorem 2. The solution of the Cauchy problem (1)-(5) is obtained as a solution of the Cauchy problem (6).

§3. Examples. Before stating our results in general form, we will give some examples to explain concretely what we have done. We will give two examples first, then we will solve them.

Example 1.

$$(7)\quad \begin{cases} rt - s^2 = 0 \\ z(0,y) = \phi(y),\ p(0,y) = \phi'(y)^2,\ q(0,y) = \phi'(y) \end{cases},$$

where $\phi(y) \in C^\infty(R^1)$. The first integral of the equation $rt-s^2=0$ are $\{p, q, z-px-qy\}$. Therefore an intermediate integral corresponding to the initial data is given by $g(x,y,z,p,q) = p - q^2$. Hence the solution of (7) is obtained as a solution of the following Cauchy problem:

$$(8)\quad \frac{\partial z}{\partial x} - \left(\frac{\partial z}{\partial y}\right)^2 = 0\ ;\qquad z(0,y) = \phi(y)\ .$$

As this is just Hamilton-Jacobi equation, we can apply the ideas of [9, 10] to construct the singularities of solutions.

Example 2.

$$(9)\quad \begin{cases} q^2 r - pqs + z(rt - s^2) = 0\ , \\ z(0,y) = \phi(y),\ \ p(0,y) = \phi(y)\ \phi'(y),\ \ q(0,y) = \phi'(y) \end{cases},$$

where $\phi(y) \in C^\infty(R^1)$. As the first integrals of (7) are $\{p, zq\}$, the intermediate integral corresponding to the initial data is $g=p-zq$. Therefore the solution of (9) is obtained as a solution of the following Cauchy problem:

$$(10)\quad \frac{\partial z}{\partial x} - z\,\frac{\partial z}{\partial y} = 0\ ;\qquad z(0,y) = \phi(y)\ .$$

As this is the conservation law, the singularities of shock type may appear.

Remark. The above examples mean that, changing the initial data, we can get intermediate integrals of various types. For example, if we give the initial data in Example 1 by $p(0,y)=-\phi'(y)$ and $q(0,y)=\phi'(y)$, then the intermediate integral is given by $g=p+q$ which is a linear equation.

First let us solve Example 1 by the characteristic method. A system of characteristic differential equations is written as

$$(11) \quad \begin{cases} \frac{dx}{d\beta} = 1 , \quad \frac{dy}{d\beta} = -2q , \quad \frac{dz}{d\beta} = -q^2 , \quad \frac{dp}{d\beta} = \frac{dq}{d\beta} = 0 , \\ x(0)=0 , \quad y(0)=\alpha , \quad z(0)=\phi(\alpha) , \quad p(0)=\phi'(\alpha)^2 , \quad q(0)=\phi'(\alpha) . \end{cases}$$

We can easily get the solutions of (11) as follows

$$(12) \quad x=\beta , \quad y=\alpha-2\phi'(\alpha)\beta , \quad z=\phi(\alpha)-\phi'(\alpha)^2\beta , \quad p=\phi'(\alpha)^2 , \quad q=\phi'(\alpha) .$$

Here we define a smooth mapping H by

$$H: R^2 \ni (\alpha,\beta) \longrightarrow (x,y)=(\beta,\alpha-2\phi'(\alpha)\beta) \in R^2 .$$

The Jacobian of H is

$$(13) \quad \frac{D(x,y)}{D(\alpha,\beta)} = 2\beta\phi''(\alpha) - 1 .$$

Therefore, as it does not vanish in a neighbourhood of $\beta=0$, we can uniquely solve the equation $H(\alpha,\beta)=(x,y)$ with respect to (α,β) in a neighbourhood of $\beta=0$ and describe them by $\alpha=\alpha(x,y)$ and $\beta=\beta(x,y)$. We define $u(x,y)=z(\alpha(x,y),\beta(x,y))$, then $u=u(x,y)$ is a classical solution of (7) in a neighbourhood of $x=0$. Here we use the notation $u(x,y)$ instead of $z(x,y)$ to avoid some confusion concerning the notations, that is to say, $u=u(x,y)$ means the solution constructed by the characteristic method.

Our problem is to consider the problem (10) in a neighbourhood of a point where the Jacobian vanishes. Here we will repeat the similar discussion as in [9, 10]. We assume the condition

$$(A.1) \quad \sup_{\alpha} \phi''(\alpha) = \phi''(\alpha_0) = M > 0 \quad \text{and} \quad \phi^{(4)}(\alpha_0) \neq 0 .$$

Then the point P where the Jacobian vanishes for the first time with respect to $x=\beta>0$ is $P=(\alpha,\beta)=(\alpha_0,1/2M)$. Let us put $H(P)=A$. From now on we consider the behavior of the solution surface $u=u(x,y)$ in a neighbourhood of the point A. The following discussion is restricted only in a neighbourhood of A.

We write $\Sigma=\{(\alpha,\beta);\ 2\beta\phi''(\alpha)-1=0\}$ and $H(\Sigma)=C$. Here we use the results of H. Whitney [11]. The assumption (A.1) assures that the point P is a cusp point of the mapping H. Therefore the point A is a cusp of the curve C. We denote by C_r the right part of C from the point A and by C_ℓ its left part from A. We denote by Ω a domain bounded by the curves C_r and C_ℓ. That is to say, the domain Ω is the image of the set $\{(\alpha,\beta);\ 2\beta\phi''(\alpha)-1\geq 0\}$ by the mapping H. Then the solution $u=u(x,y)$ becomes three-valued on the domain Ω. Let us analyse this situation more precisely. The discussion of this part are essentially same to one of Tsuji [9, 10]. It is easy to extend the classical solution $u=u(x,y)$ from $x=0$ to $x=1/2M$. Considering the structure of the mapping H in the neighbourhood of Σ, we can see that the solution defined in the domain of the left side from the point A, i.e., $u(x,y)$ defined on $\{x=1/2M,\ y<y(\alpha_0,1/2M)\}$, can be extended up to the curve C_r, and that it is reflected along C_r. The solution defined in the right side from A can be also extended to the curve C_ℓ and it is reflected along C_ℓ. The former reflected solution along C_r coincides with the latter reflected one along C_ℓ. See Figure 1.

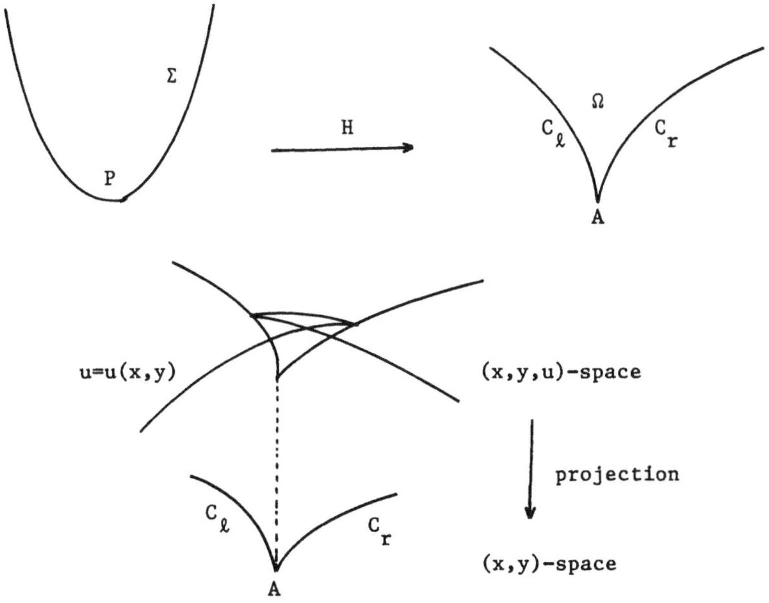

Figure 1

Moreover we can see by [10] that the solution is of class C^1 up to the boundary C_r and also to C_ℓ, but it is not of C^2 in a neighbourhood of C. We denote $(u,\partial u/\partial x,\partial u/\partial y)\big|_{C_r}=(u_r,p_r,q_r)$ and $(u,\partial u/\partial x,\partial u/\partial y)\big|_{C_\ell}=(x_\ell,p_\ell,q_\ell)$. In this case we may expect that, though the solution $u=u(x,y)$ is reflected along C_r and C_ℓ, it would be possible to extend the solution beyond C_r and C_ℓ to another side of Ω in the class C^1. But we can show that it is impossible. Let us pick up any point R on C_r and U be any small open neighbourhood of R. Put $\Omega_r=U-U\cap\Omega$. Consider the Cauchy problem

$$(14)\quad \frac{\partial z}{\partial x}-\left(\frac{\partial z}{\partial y}\right)^2=0 \quad \text{in } \Omega_r \ ; \quad \left(z,\frac{\partial z}{\partial x},\frac{\partial z}{\partial y}\right)=(u_r,p_r,q_r) \quad \text{on} \quad C_r\cap U.$$

Proposition 3. The Cauchy problem (14) has not a solution of class C^1 for any open neighbourhood U of the point R.

A. Plis [8] proved that every C^1-solution of first order partial differential equation is generated by characteristic strips. We use this result to prove the above proposition. Proposition 3 means that we can not extend the solution $u=u(x,y)$ beyond the curves C_r and C_ℓ in C^1-space, that is to say, that there is no other solution than $u(x,y)$ in a neighbourhood of C. Summing up the above results, we have the following

Theorem 4. Suppose the condition (A.1). Concerning the Cauchy problem (7), it holds

(i) The problem (7) has uniquely a C^∞-solution in $\{0\le x<1/2M,\ y\in R^1\}$,

(ii) The solution $u=u(x,y)$ is three-valued on Ω, and there exists an open neighbourhood U of the point A such that $u(x,y)\in C^\infty(U-C)\cap C^1(U)$ but $u(x,y)\notin C^2(U)$,

(iii) The problem (7) has no other solution than $u(x,y)$ in the space of functions which are of class C^1 and piecewise C^2.

Next we advance to Example 2. Characteristic equations for (10) are written as

$$\frac{dx}{d\beta}=1\ ,\quad \frac{dy}{d\beta}=-z\ ,\quad \frac{dz}{d\beta}=0\quad ;\quad x(0)=0\ ,\ y(0)=\alpha\ ,\ z(0)=\phi(\alpha).$$

The solutions of this system are easily obtained as $\{x=\beta,\ y=\alpha-\beta\phi(\alpha),\ z=\phi(\alpha)\}$. As in Example 1, we define a smooth mapping H as

$$H : R^2 \ni (\alpha,\beta) \longrightarrow (x,y)=(\beta,\alpha-\beta\phi(\alpha)) \in R^2.$$

By the similar reasoning to Example 1, we get a smooth solution in a neighbourhood of $x=0$. As a hypothesis corresponding to (A.1), we put the condition:

$$\text{(A.2)} \qquad \sup_{\alpha} \phi'(\alpha) = \phi'(\alpha_0) = M > 0 \quad \text{and} \quad \phi^{(3)}(\alpha_0) \neq 0 .$$

We use the same notations as in Example 1. For example, we write $\Sigma=\{(\alpha,\beta);\ D(x,y)/D(\alpha,\beta)=\beta\phi'(\alpha)-1=0\}$, $H(\Sigma)=C$, etc. Though the behavior of the solution surface of Example 2 is similar to Example 1, there exist several differences. The principal one is that the solution of Example 2 is continuous up to the boundary C, but that it is not of C^1 in a neighbourhood of C. Moreover Proposition 3 must be modified a little.
The problem in Example 2 which is corresponding to (14) in Example 1 is expressed as follows:

$$\text{(15)} \qquad \frac{\partial z}{\partial x} - z\frac{\partial z}{\partial y} = 0 \quad \text{in} \quad \Omega_r \ ; \quad z=u_r \quad \text{on} \quad C_r \cap U.$$

Proposition 5. The Cauchy problem (15) has not a solution of class C^1 for any open neighbourhood U of the point R.

Therefore we can not extend the smooth solution of (9) beyond the curve C. Summing up these results, we have the following

Theorem 6. Consider the Cauchy problem (9) under the condition (A.2), then it holds

(i) The Cauchy problem (7) has uniquely a C^∞-solution in $\{0\leq x<1/M,\ y\in R^1\}$,

(ii) The solution $u=u(x,y)$ is three-valued on Ω, and there exists an open neighbourhood U of A such that $u(x,y) \in C^\infty(U-C) \cap C^0(U)$, but $u(x,y) \notin C^1(U)$,

(iii) The Cauchy problem (7) has no other solution than $u(x,y)$ in the space of functions which are of $C^0(U)$ and piecewise C^1.

Remark. The geometrical point of view, a solution surface is regarded as a contact manifold. Theorem 4 shows that this approach is appropriate for Example 1. But, for Example 2, this point of view must be modified a little, because $p(\alpha,\beta)$ and $q(\alpha,\beta)$ can not be defined for $(\alpha_0,\beta_0) \in \Sigma$. See Tsuji [10]. Therefore the solution surface may loose the structure of contact manifold on Σ.

§4. Generalization. Let us generalize the results of Example 1 and Example 2. We consider the Cauchy problem for Monge-Ampère equation (1)-(5) which satisfies the assumptions (H1), (H2) and (H3). Then we can reduce the problem (1)-(5) to the Cauchy problem (6) obtained by intermediate integral. We describe the hamiltonian flow for (6) by $x=x(\alpha,\beta)$, $y=y(\alpha,\beta)$, $z=z(\alpha,\beta)$, $p=p(\alpha,\beta)$ and $q=q(\alpha,\beta)$, and define a smooth mapping H by

$$H: R^2 \ni (\alpha,\beta) \longrightarrow (x,y)=(x(\alpha,\beta),\ y(\alpha,\beta)) \in R^2.$$

We assume the following condition:

(H4) The singularities of the mapping H are fold and cusp points only.

H. Whitney [11] suggests us that this assumption is natural in generic sense. If the equation (6) is a quasi-linear equation of first order, $p(\alpha,\beta)$ and $q(\alpha,\beta)$ can not be defined on the singularities of H which is a set $\Sigma=\{(\alpha,\beta);\ D(x,y)/D(\alpha,\beta)=0\}$. For example, if $(\alpha_0,\beta_0)\in\Sigma$, $p(\alpha,\beta)$ and $q(\alpha,\beta)$ tend to infinity when (α,β) $((\alpha,\beta)\notin\Sigma)$ tends to (α_0,β_0). But, for classical Hamilton-Jacobi equations, $p(\alpha,\beta)$ and $q(\alpha,\beta)$ can exist even if the Jacobian of the mapping H may vanish. This is the principal difference between quasi-linear equations and Hamilton-Jacobi equations. This fact leads us to call first order partial differential equations as "Hamilton-Jacobi type" when characteristic strips exist even if the mapping H may have singularities. More precisely, see [10]. As the generalization of Example 1, we have the following

Theorem 7. Assume the conditions (H1), (H2), (H3) and (H4). Moreover assume that the equation (6) is of Hamilton-Jacobi type. Then, in a neighbourhood of singular points of H, the behavior of the solution surface is same to the one of Example 1.

As the generalization of Example 2, we get

Theorem 8. Assume the conditions (H1), (H2), (H3) and (H4). Moreover assume that the equation (6) is a quasi-linear partial differntial equation of first order. Then, in a neighbourhood of singular points of H, the behavior of the solution surface is same to the one of Example 2.

REFERENCES

[1] R. Courant and D. Hilbert, Method of Mathematical Physics, vol. 2, Interscience, New York (1962).

[2] G. Darboux, Leçons sur la thorie générale des surfaces, tome 3, Gauthier-Villars, Paris (1894).

[3] E. Goursat, Leçons sur l'intégration des équations aux dérivées partielles du second ordre, tome 1, Hermann, Paris (1896).

[4] ———— , Cours d'analyse mathématiques, tome 3, Gauthier-Villars, Paris (1927).

[5] J. Hadamard, Le problème de Cauchy et les équations aux dérivées partielles lin aires hyperboliques, Hermann, Paris (1932).

[6] H. Lewy, Über das Anfangswertproblem bei einer hyperbolischen nichtlinearen partiellen Differentialgleichung zweiter Ordnung mit zwei unabhängigen Veränderlichen, Math. Ann., 98, 179-191 (1928).

[7] ————, A priori limitations for solutions of Monge-Ampère equations I, II. Trans. A. M. S., 37, 417-434 (1935); 41, 365-374.

[8] A. Pliš, Characteristics of nonlinear partial differential equations, Bull. Acad. Polon. Sci., Cl. III, 2, 419-422 (1954).

[9] M. Tsuji, Formation of singularities for Hamilton-Jacobi equation II, J. Math. Kyoto Univ., 26, 299-308 (1986).

[10] ————, Prolongation of classical solutions and singularities of generalized solutions, Ann. Inst. H. Poincaré -Analyse Nonlinéaires, 7, 505-523 (1990).

[11] H. Whitney, On singularities of mappings of Euclidean spaces I, Ann. Math., 62, 374-410 (1955).

Third International Colloquium on Differential Equations pp. 205-217 (1993)
D. Bainov and V. Covachev (Eds)

Oscillation properties of solutions of arbitrary order neutral differential equations

A. Zafer and R. S. Dahiya

Department of Mathematics
Iowa State University
Ames, IA 50010

Abstract

In this paper we study the oscillation behavior of solutions of neutral functional differential equation of the form

$$\frac{d}{dt}[a(t)[x(t)+p(t)x(\tau(t))]^{(n-1)}]+q(t)f(x(\sigma(t)))=0. \tag{1}$$

Sufficient conditions are provided for the above differential equation to be almost oscillatory in the sense that every solution $x(t)$ of (1) is either oscillatory or else $\lim_{t\to\infty}|x(t)|=\infty$, or $\liminf_{t\to\infty}|x(t)|=0$.

1 Introduction

Although the oscillation theory of functional differential equations with deviating arguments has been developed extensively in the last 20 or 30 years, that of neutral functional differential equations, i.e., equations in which the highest order derivative of the unknown function appear both with and without deviating arguments, was not investigated in the same respect. In recent years, however, the oscillation of neutral functional differential equations are studied the most. In the following we state some known oscillation criteria for a neutral differential equation. For more results, we refer the reader to [1, 10, 11, 12, 13, 14, 19].

Let

$$(x(t)+p(t)x(t-\tau))''+q(t)x(t-\sigma)=0,$$

where $p\in C([t_0,\infty)), 0\le p(t)<1$; $q\in C([t_0,\infty))$, $q(t)$ is not identically zero in any half-line of the form $[t,\infty)$; and the delays τ and σ are nonnegative.

This equation was first studied by J. R. Graef, et al. [11]. They proved that if

$$\int^{\infty} q(t)(1-p(t-\sigma))\,dt=\infty,$$

then every solution $x(t)$ is oscillatory.

Later, S. R. Grace and B. S. Lalli [10] considered the more general equation

$$(a(t)(x(t)+p(t)x(t-\tau))')'+q(t)f(x(t-\sigma))=0,$$

where p, q, σ, and τ are as above; $a \in C([t_0,\infty))$, $a(t)>0$; $f \in C((-\infty,\infty))$, $xf(x) \geq cx^2$ for $x \neq 0$ for some $c>0$. They showed that if there exists a function $\rho \in C([t_0,\infty), R^+)$ such that

$$\int^\infty [\rho(s)q(s)(1-p(s-\sigma))-\frac{\rho'(s)^2a(s-\sigma)}{4\rho(s)}]\,ds=\infty,$$

then every solution $x(t)$ is oscillatory.

Recently, A. S. Tantawy and R. S. Dahiya [21] proved that the results of J. R. Graef, et al. remains true if the second order derivative is replaced by an arbitrary even order derivative.

Our purpose in this paper is to study the oscillatory behavior of solutions of n-th order neutral functional differential equations of the form

$$[a(t)[x(t)+p(t)x(\tau(t))]^{(n-1)}]'+\delta q(t)f(x(\sigma(t)))=0, \quad \delta=\pm 1. \tag{2}$$

and generalize several results obtained for second and n-th order equations of much simpler type.

In what follows, by a *proper* solution of (2) we mean a function $x:[T_x,\infty)\to R$ which satisfies (2) for sufficiently large t, and $\sup\{|x(t)| : t \geq T\}>0$ for any $T \geq T_x$. A proper solution of (2) is called oscillatory if it has arbitrarily large zeros; otherwise it is called nonoscillatory. Thus a nonoscillatory solution is necessarily eventually positive or eventually negative. We make the standing hypothesis that (2) posses proper solutions. For questions on existence, uniqueness, and continuous dependence, see Driver [6], Bellman and Cooke [2], and Hale [17].

Throughout unless we state otherwise, the following conditions are always assumed to hold:

(a) $a \in C([t_0,\infty),R)$, $a(t)>0$, and

$$\int^\infty \frac{1}{a(t)}\,dt=\infty; \tag{3}$$

(b) $p \in C([t_0,\infty),R)$, $0 \leq p(t) < 1$;

(c) $q \in C([t_0,\infty),R^)$, $q(t)>0$;

(d) $f \in C(R,R)$ and $xf(x)>0$ for $x \neq 0$;

(e) $\tau(t) \leq t$, $\tau(t)$ and $\sigma(t) \to \infty$ as $t \to \infty$.

2 Lemmas

Here we present some lemmas which we will rely on in the next section.

Lemma 2.1 (Kiguradze [18]) *If $u(t)$ is an n-times differentiable function on R^+ of constant sign, $u^{(n)}(t)$ is of constant sign and not identically zero in any interval $[t_1, \infty)$, and*

$$u(t)u^{(k)}(t) \leq 0,$$

then
(i) there exists a $t_2 \geq t_1$ such that the functions $u^{(k)}(t)$, $k = 1, 2, ..., n-1$, are of constant sign on $[t_1, \infty)$,
(ii) there is an integer l, $0 \leq l \leq n-1$ with $n-l$ is odd, such that for $t \geq t_2$

$$\begin{aligned} u(t)u^{(k)}(t) &> 0 \quad k = 0, 1, ..., l \\ (-1)^{n-k-1}u(t)u^{(k)}(t) &> 0 \quad k = l, ..., n-1, \end{aligned} \tag{4}$$

and if $1 \leq l \leq n-1$,

$$|u^{(l-k-1)}(t)| \geq \frac{t - t_2}{k+1}|u^{(l-k)}(t)| \quad for\ k = 0, 1, ..., l-1. \tag{5}$$

If $u(t)$ satisfies (4), then it is said to be a nonoscillatory function of degree l [9].

Lemma 2.2 (Kiguradze [18]) *Let $u(t)$ be an n-times differentiable function on $[t_1, \infty)$ with $u^{(k)}(t)$, $k = 0, ..., n-1$, absolutely continuous and of constant sign on$[t_1, \infty)$, and let $u(t)u^{(k)}(t) \geq 0$ for every t in $[t_1, \infty)$, then either*

$$u(t)u^{(k)}(t) > 0 \; k = 0, 1, 2, ..., n-1 \tag{6}$$

or there is an integer l, $0 \leq l \leq n-2$ with $n-l$ is even, such that for $t \geq t_1$

$$\begin{aligned} u(t)u^{(k)}(t) &> 0 \quad k = 0, 1, ..., l \\ (-1)^{n-k}u(t)u^{(k)}(t) &> 0 \quad k = l, ..., n-1, \end{aligned} \tag{7}$$

and if $1 \leq l \leq n-1$ then inequality (5) holds.

Lemma 2.3 (Q. Chuanxi and G. Ladas [4]) *Let $z, x, p : [t_0, \infty) \to R$ and $c \in R$ be such that*

$$z(t) = x(t) + p(t)x(t-c), \quad t \geq t_1 = t_0 + \max\{t_0, c\}$$

Assume that there exist real numbers p_1, p_2, p_3, p_4 such that $p(t)$ is in one of the following ranges:

$$\begin{aligned} p_1 &\leq p(t) \leq 0; \\ 0 &\leq p(t) \leq p_2 < 1; \\ 1 &\leq p_3 \leq p(t) \leq p_4. \end{aligned}$$

Suppose that

$$x(t) > 0 \quad for\ t \geq t_0, \ \liminf_{t\to\infty} x(t) = 0$$

and that

$$\lim_{t\to\infty} z(t) = L \in R \ exists.$$

Then $L = 0$.

It should be noted that the solutions of (2) need not be continuously differentiable. All we need is that $z(t) = x(t) + p(t)x(\tau(t))$ together with its derivatives up to order n be continuous. The following lemma contains more about this function $z(t)$.

Lemma 2.4 *Let $x(t)$ be a nonoscillatory solution of (1.1) and let $z(t)$ be a function defined by $z(t) = x(t) + p(t)x(\tau(t))$.*

(i) There exist a $T > 0$ such that for $\delta = 1$

$$z^{(n-1)}(t) > 0, \quad t \geq T,$$

and for $\delta = -1$

$$\text{either } z^{(n-1)}(t) < 0, \quad t \geq T, \text{ or } \lim_{t\to\infty} z^{(n-2)}(t) = \infty.$$

(ii) If $a'(t) \geq 0$, then there exists an integer l, $l \in \{0, 1, ..., n\}$ with $(-1)^{n-l-1}\delta = 1$, such that for $t \geq T$

$$\begin{aligned} z(t)z^{(k)}(t) &> 0 \quad k = 0, 1, ..., l \\ (-1)^{k-l}z(t)z^{(k)}(t) &> 0 \quad k = l, ..., n, \end{aligned} \tag{8}$$

and if $1 \leq l \leq n-1$ then

$$|z^{(l-k-1)}(t)| \geq \frac{t-T}{k+1}|z^{(l-k)}(t)| \quad \text{for } k = 0, 1, ..., l-1 \tag{9}$$

Proof: We may assume that $x(t)$, $x(\tau(t))$, and $x(\sigma(t))$ are positive for all $t \geq t_1 \geq t_0$. The proof when $x(t)$ is eventually negative can be constructed similarly. Clearly

$$z(t) > 0 \text{ for } t \geq t_1, \tag{10}$$

and from (1.1)

$$\delta(a(t)z^{(n-1)}(t))' = -q(t)f(x(\sigma(t))) \leq 0.$$

Since $\delta a(t)z^{(n-1)}(t)$ is a decreasing function for $t \geq t_1$, we can have

$$\delta a(t)z^{(n-1)}(t) > 0 \text{ for } t \geq t_1 \tag{11}$$

or

$$\delta a(t)z^{(n-1)}(t) < 0 \text{ for } t \geq T_1 \geq t_1. \tag{12}$$

Suppose that (12) holds. Then

$$\delta a(t)z^{(n-1)}(t) \leq \delta a(T_1)z^{(n-1)}(T_1) < 0 \quad \text{for } t \geq T_1.$$

Integrating the above inequality divided by $a(t)$ from T_1 to t and using (3) we have

$$\delta z^{(n-2)}(t) \to -\infty \text{ as } t \to \infty.$$

Since this implies that $\delta z(t) \to -\infty$ as $t \to \infty$, in view of (10) we easily conclude that $\delta = -1$. This completes the proof of part (i) of the lemma.

To prove (ii) we first notice that, since

$$\delta[a(t)z^{(n-1)}(t)]' = \delta a(t)z^{(n)}(t) + \delta a'(t)z^{(n-1)}(t) = -q(t)f(x(\sigma(t))),$$

we have

$$\delta z^{(n)}(t) = -\delta\frac{a'(t)}{a(t)}z^{(n-1)}(t) - \frac{q(t)}{a(t)}f(x(\sigma(t))). \tag{13}$$

Suppose that $\lim_{t\to\infty} z^{(n-2)}(t) \neq \infty$ when $\delta = -1$. Then it follows from (i) and (13) that

$$\delta z^{(n}(t) < 0. \tag{14}$$

In view of (10) and (14), applying Lemma 2.1 and Lemma 2.2 it follows that there exists a $T \geq t_1$ and an integer l, $0 \leq l \leq n$ with $(-1)^{n-l-1}\delta = 1$, such that for $t \geq T$,

$$\begin{aligned} z^{(k)}(t) &> 0, \ k = 0, 1, 2, ..., l \\ (-1)^{k-l}z^{(k)}(t) &> 0, \ k = l, l+1, ..., n, \end{aligned} \tag{15}$$

and (9) holds when $1 \leq l \leq n-1$.

Suppose that $\lim_{t\to\infty} z^{(n-2)}(t) = \infty$ and $\delta = -1$. Then, $z^{(n-1)}(t)$ is eventually positive and so from (13) $z^{(n)}(t)$ is also eventually positive. It follows that $z^{(i)}$, $i = 0, 1, ..., n$ are all eventually positive. But, this case is included in (15).

3 Main results

Eq. (2) is said to be of retarted type if $\sigma(t) < t$, mixed type if $\sigma(t) > t$. In the following theorems we do not impose any condition on the argument $\sigma(t)$, so the results proved here are quite general.

Theorem 3.1 *Assume that f is increasing and*

$$\int^{\infty} q(s)f((1-p(\sigma(s))c)\,ds = \infty, \tag{16}$$

for all $c > 0$.

(i) If $\delta = 1$, then every solution $x(t)$ of (2) is oscillatory when n is even, and every solution $x(t)$ of (2) is either oscillatory or satisfies

$$\liminf_{t\to\infty} |x(t)| = 0$$

when n is odd .

(ii) If $\delta = -1$, then every solution $x(t)$ of (2) is either oscillatory or else

$$\lim_{t\to\infty} |x(t)| = \infty \quad \text{or } \liminf_{t\to\infty} |x(t)| = 0$$

when n is even, and every solution $x(t)$ of (2) is either oscillatory or else

$$\lim_{t\to\infty} |x(t)| = \infty$$

when n is odd.

Proof : Let $x(t)$ be a nonoscillatory solution of (2). We may assume that eventually $x(t) > 0$. The case $x(t) < 0$ can be treated similarly. By the part (i) of Lemma 2.4, we see that there exist a $t_1 > 0$ such that for $\delta = 1$

$$z^{(n-1)}(t) > 0 \tag{17}$$

and for $\delta = -1$, either

$$z^{(n-1)}(t) < 0 \tag{18}$$

or

$$\lim_{t\to\infty} z^{(n-2)}(t) = \infty \tag{19}$$

Suppose that $\lim_{t\to\infty} z^{(n-2)}(t) \neq \infty$ when $\delta = -1$. Then, in view of (17) and (18), it follows from Lemma 2.1 and Lemma 2.2 that there exist a $T \geq t_1$ and an integer $l \in \{0, 1, ..., n-1\}$ with $(-1)^{n-l-1}\delta = 1$, such that for $t \geq T$

$$\begin{aligned} z^{(i)}(t) &> 0, \; i = 0, 1, 2, ..., l \\ (-1)^{i-l} z^{(i)}(t) &> 0, \; i = l, l+1, ..., n-1. \end{aligned} \tag{20}$$

Let n be even and $\delta = 1$, or n be odd and $\delta = -1$. It is clear from (20) that $z(t)$ is increasing. Therefore, in view of $x(t) \leq z(t)$, we have for $t \geq t_2$,

$$\begin{aligned} z(t) &\leq x(t) + p(t)z(\tau(t)) \\ &\leq x(t) + p(t)z(t) \end{aligned}$$

or

$$(1 - p(t))z(t) \leq x(t). \tag{21}$$

On the other hand, $z(t) > 0$ and increasing and $\sigma(t) \to \infty$ as $t \to \infty$ imply that there exist a $c > 0$ and a $t_3 \geq t_2$ such that

$$z(\sigma(t)) \geq c \text{ for } t \geq t_3. \tag{22}$$

Now, we integrate (2) from t_3 to t to obtain

$$\delta a(t) z^{(n-1)}(t) - \delta a(t_3) z^{(n-1)}(t_3) + \int_{t_3}^{t} q(s) f(x(\sigma(s)))\, ds = 0. \tag{23}$$

By (21), (22) and the fact that f is increasing, we have

$$f(x(\sigma(t))) \geq f((1 - p(\sigma(t)))c) \text{ for } t \geq t_3. \tag{24}$$

Using (24) in (23) leads to

$$\delta a(t) z^{(n-1)}(t) - \delta a(t_3) z^{(n-1)}(t_3) + \int_{t_3}^{t} q(s) f((1 - p(\sigma(s)))c))\, ds \leq 0. \tag{25}$$

In view of (16), we conclude from (25) that

$$\delta a(t) z^{(n-1)}(t) \to -\infty \text{ as } t \to \infty. \tag{26}$$

But (26) is a contradiction with (17) and (18). This proves that $x(t)$ is oscillatory when $\delta = 1$, and $x(t)$ is either oscillatory or (19) holds when $\delta = -1$. Clearly if (19) holds, then $\lim_{t\to\infty} x(t) = \infty$.

Let n be odd and $\delta = 1$, or n be even and $\delta = -1$. If the integer l associated with $z(t)$ is positive, then we can easily arrive at the above conclusion. Thus, l must be zero. Noting that

$$\int^{\infty} q(t)\, dt = \infty$$

and

$$\lim_{t\to\infty} \delta a(t) z^{(n-1)}(t) = L \geq 0,$$

it is not difficult to see from (23) and (??) that

$$\liminf_{t\to\infty} f(x(t)) = 0 \text{ or } \liminf_{t\to\infty} x(t) = 0.$$

This clearly completes the proof of Theorem 3.1.

Corollary 3.1 *Let $(-1)^n\delta = -1$, $\tau(t) = t - \tau$, and suppose that there exists a positive number p which satisfies $0 \leq p(t) \leq p < 1$. Then, every solution $x(t)$ of (2) is either oscillatory or satisfies $\lim_{t\to\infty} x(t) = 0$ when n is odd, and every solution $x(t)$ of (2) is either oscillatory or else $\lim_{t\to\infty} |x(t)| = \infty$ or $\lim_{t\to\infty} x(t) = 0$ when n is even.*

Proof: Let $x(t)$ be an eventually positive solution of (2), and let $l = 0$. Since then $z(t) > 0$ and $z'(t) < 0$ for t sufficiently large, $\lim_{t\to\infty} z(t)$ exists. Applying lemma 2.3, we obtain that

$$\lim_{t\to\infty} z(t) = 0.$$

On the other hand, $0 < x(t) \leq z(t)$. Therefore

$$\lim_{t\to\infty} x(t) = 0$$

as desired.

Example 1. Consider the equation

$$[e^{-t}[x(t) + 2e^{\pi-t}x(t-\pi)]''']' + 2\sqrt{2}e^{7\pi/4}(1+e^{-t})x(t-7\pi/4) = 0 \qquad (27)$$

so that $\delta = 1$, $n = 4$, $a(t) = e^{-t}$, $p(t) = 2e^{3\pi/2-t}$, $\tau(t) = t - \pi/2$, $\sigma(t) = t - 7\pi$, $q(t) = 2\sqrt{2}e^{7\pi/4}(1+e^{-t})$. According to Theorem 3.1, every solution of (27) is oscillatory. In fact, $x(\ ) = \sqrt{2}e^t \cos(t - \pi/4)$ is a solution of (27).

Example 2. Consider the equation

$$[\frac{1}{t}[x(t) + \frac{4}{2t-3\pi}x(t-3\pi/2)]'']' - \frac{2}{2t-\pi}x(t-\pi/2) = 0 \qquad (28)$$

so that $\delta = -1$, $n = 3$, $a(t) = 1/t$, $p(t) = 4/(2t-3\pi)$, $\tau(t) = t - \pi/2$, $\sigma(t) = t - 7\pi$, $q(t) = 2/(2t-\pi)$. According to Theorem 3.1, every solution of (28) is either oscillatory or tends to infinity as t approaches infinity. It is easy to verify that $x(t) = t\cos t$ is a solution of (28).

Example 3. Consider the equation

$$[x(t) + e^{\pi-t}x(t-\pi)]^{(4)} + e^{\pm 2\pi}(4+e^{-t})x(t \pm 2\pi) = 0 \qquad (29)$$

so that $\delta = 1$, $n = 4$, $a(t) = 1$, $p(t) = e^{\pi/2-t}$, $\tau(t) = t - \pi/2$, $\sigma(t) = t \pm 2\pi$, $q(t) = e^{2\pi}(4 + e^{-t})$. According to Theorem 3.1, every solution of (30) is oscillatory. In fact, $x(t) = e^t \sin t$ is a solution of (30).

Remarks:

1. If $\delta = 1$, $n = 2$, $a(t) = 1$, $\tau(t) = t - \tau$, $\sigma(t) = t - \sigma$, and $f(x) = x$, then the results of J. R. Graef et al.[11] is obtained.

2. If $\delta = 1$, n is even, $\tau(t) = t - \tau$, $\sigma(t) = t - \sigma$, $a(t) = 1$ and $f(x) = x$, we recover the theorem proved by A. S. Tantawy and R. S. Dahiya [21].

3. If $\delta = 1$, $n = 2$, $\tau(t) = t - \tau$, $\sigma(t) = t - \sigma$, and $xf(x) > cx^2 > 0$ for $x \neq 0$, we have the results of B. S. Lalli and S. R. Grace [10] for $\rho(t) = 1$.

If we restrict ourselves to the bounded solutions of (2), then the condition (16) of Theorem 3.1 can be weakened as follows:

Theorem 3.2 *Let f be increasing and $a(t) = 1$. Suppose that*

$$\int^{\infty} t^{n-1} q(t) f(1 - p(\sigma(t))c)\, dt = \infty \tag{31}$$

for every $c > 0$. Then

(i) every solution $x(t)$ of (2) is oscillatory when $(-1)^n\delta = 1$, and
(ii) every solution of $x(t)$ of (2) is either oscillatory or satisfies

$$\liminf_{t\to\infty} |x(t)| = 0$$

when $(-1)^n\delta = -1$.

Proof: Let $x(t)$ be a nonoscillatory solution of (2). We may assume that $x(t) > 0$ eventually. From (e), we see that there is a $t_1 \geq t_0$ such that $x(t)$, $x(\tau(t))$, and $x(\sigma(t))$ are positive for $t \geq t_1$. Setting $z(t) = x(t) + p(t)x(\tau(t))$, it can be seen from (2) that $\delta z^{(n)}(t) \leq 0$ for $t \geq t_1$. In view of (c), we obtain that $z^{(n-1)}(t)$ is decreasing and that the derivatives of $z(t)$ up to $n - 1$ are eventually of constant sign. Since $z(t)$ is bounded it follows that there exists a $t_2 \geq t_1$ such that for $t \geq t_2$,

$$(-1)^{k-1}\delta z^{(k)}(t) > 0 \ \text{ for } k = 1, 2, ..., n-1 \quad \text{if } n \text{ is even} \tag{32}$$

and

$$(-1)^{k}\delta z^{(k)}(t) > 0 \ \text{ for } k = 1, 2, ..., n-1 \quad \text{if } n \text{ is odd.} \tag{33}$$

In both cases, we have

$$\lim_{t\to\infty} z^{(k)}(t) = 0 \ \text{ for } k = 1, 2, ..., n-1. \tag{34}$$

Using (34), we now integrate (2) n-times between t and ∞ to obtain

$$(-1)^n\delta[z(\infty) - z(t)] = \frac{1}{(n-1)!}\int_t^{\infty} (s-t)^{n-1} q(s) f(x(\sigma(s)))\, ds, \tag{35}$$

where $z(\infty) = \lim_{t\to\infty} z(t)$.

If $(-1)^n\delta = 1$, then we see from (32) and (33) that $z(t)$ is increasing for t large. Since $z(t)$ is also positive, we have as in Theorem 3.1,

$$f(x(\sigma(t))) \geq f((1-p(\sigma(t))c) \tag{36}$$

for $t \geq t_2$ for some $t_2 \geq t_1$ and $c > 0$. Thus, from (35) and (36) we get

$$z(\infty) - z(t_2) \geq \frac{1}{(n-1)!}\int_{t_2}^{\infty}(s-t_2)^{n-1}q(s)f((1-p(\sigma(s))c))\,ds, \tag{37}$$

which, because of (31), implies that $z(\infty) = \infty$ and so contradicts to boundedness of $z(t)$.

If $(-1)^n\delta =$, then we see that $z(t)$ is positive and decreasing for $t \geq t_2$. So, from (35) it follows that

$$z(t_2) \geq \frac{1}{(n-1)!}\int_{t_2}^{\infty}(s-t_2)^{n-1}q(s)f((1-p(\sigma(s))c))\,ds, \tag{38}$$

On the other hand, since

$$\int^{\infty} t^{n-1}q(t)\,dt = \infty,$$

it follows from (38) that

$$\liminf_{t\to\infty} f(x(t)) = 0 \text{ or } \liminf_{t\to\infty} x(t) = 0.$$

This completes the proof of Theorem 3.2 when $x(t)$ is eventually positive. The proof when $x(t)$ is eventually negative is similar.

If $a(t) \neq 1$, then the previous theorem can be generalized provided $a'(t) \geq 0$. Specifically, we shall prove

Theorem 3.3 *Let $a'(t) \geq 0$ for $t \geq t_0$, and let f be increasing. If*

$$\int^{\infty}\frac{1}{a(t)}t^{n-1}q(t)f(1-p(\sigma(t))c)\,ds = \infty \tag{39}$$

for every $c > 0$, then the conclusion of the theorem 3.2 holds.

Proof: Suppose that there exists an eventually positive bounded solution of (2). Let $z(t) = x(t) + p(t)x(\tau(t))$. Then by part (ii) of Lemma 2.4, there is a $T > 0$ and an integer $l \in \{0, 1\}$ with $(-1)^{n-l-1}\delta = 1$, such that for $t \geq T$

$$\begin{aligned} z^{(k)}(t) &> 0 \quad k = 1, 2, \ldots l \\ (-1)^{n-k-1}z^{(k)}(t) &> 0 \quad k = l, \ldots n-1. \end{aligned} \tag{40}$$

By Taylor's formula for $r \geq t \geq T$,

$$z^{(l)}(t) = \sum_{i=0}^{n-l-1}(-1)^i\frac{z^{(l+i)}(r)}{(i)!}(r-t)^i + \int_t^r \frac{(t-r)^{n-l-1}}{(n-l-1)!}(-z^{(n)}(r))\,dr \tag{41}$$

In view of (40), it follows from (41) that

$$z^{(l)}(t) \geq \frac{1}{(n-l-1)!}\int_t^r (s-t)^{n-l-1}(-z^{(n)}(s))\,ds \text{ for } T \leq t < r.$$

Letting $r \to \infty$ and using (13), thus we obtain

$$z^{(l)}(t) \geq \frac{1}{(n-l-1)!}\int_t^\infty (s-t)^{n-l-1}[\frac{q(s)}{a(s)}f(x(\sigma(s))) + \frac{a'(s)}{a(s)}z^{(n-1)}(s)]\,ds$$

and hence

$$z^{(l)}(t) \geq \frac{1}{(n-l-1)!}\int_t^\infty (s-t)^{n-l-1}\frac{q(s)}{a(s)}f(x(\sigma(s)))\,ds. \tag{42}$$

Clearly if $(-1)^n\delta = -1$, then $l = 0$, and if $(-1)^n\delta = 1$, then $l = 1$. In both cases, the remainder of the proof is similar to that of theorem 3.3, and so it is omitted.

The proof of the following corollary is exactly the same as that of Corollary 3.2.

Corollary 3.2 *Let $(-1)^n\delta = -1$, $\tau(t) = t - \tau$, and suppose that there exists a positive number p which satisfies $0 \leq p(t) \leq p < 1$. Then*

(i) if $\delta = 1$, every solution $x(t)$ is either oscillatory or satisfies $\lim_{t\to\infty} x(t) = 0$, and
(ii) if $\delta = -1$, every solution $x(t)$ is either oscillatory or $\lim_{t\to\infty} x(t) = 0$.

Example 4. Consider the equation

$$[t^{-2/3}[x(t) + \frac{t-1}{2t}x(t-1)]''']' - 42(t-2)^{5/3}t^{-17/3}x(t-2)^{5/3} = 0 \tag{43}$$

so that $\delta = -1$, $n = 4$, $a(t) = t^{-2/3}$, $p(t) = \frac{t-1}{2t}$, $\tau(t) = t-1$, $\sigma(t) = t-2$, $q(t) = 42(t-2)^{5/3}t^{-17/3}$, and $f(x) = x^{5/3}$. We note that Theorem 3.1 cannot be applied here, since (16) does not hold. However, by Corollary 3.2, every solution of (43) is either oscillatory or tends to zero as t approaches infinity. It is easy to verify that $x(t) = 1/t$ is a solution of (43).

We conclude this paper with an oscillation therem for all solutions of (2) in the case when $a(t)$ and $q(t)$ are periodic functions.

Before we state and prove the result, we note that

$$\int^\infty \frac{1}{a(t)}\,dt = \int^\infty q(t)\,dt = \infty,$$

since $a(t)$ and $q(t)$ are continuous and periodic.

Theorem 3.4 *Suppose that $p(t) = p > 0$, $\tau(t) = t - \tau$, $\sigma(t) = t - \sigma$, $q(t)$ and $a(t)$ are τ periodic, and that f is increasing and satisfies*

$$\begin{aligned} f(x+y) &\leq f(x) + f(y), \quad x, y > 0, \\ f(x+y) &\geq f(x) + f(y), \quad x, y < 0, \\ f(\lambda x) &\leq \lambda f(x), \quad x, \lambda > 0, \\ f(\lambda x) &\geq \lambda f(x), \quad x < 0, \lambda > 0. \end{aligned}$$

Then the conclusion of the theorem 3.1 holds.

Proof: Let $x(t)$ be a nonoscillatory solution of (2), say $x(t) > 0$ for $t \geq t_1$. Set $z(t) = x(t) + px(t-\tau)$, $w(t) = z(t) + pz(t-\tau)$, $v(t) = w(t) + pw(t-\tau)$. Then clearly there exists $t_2 \geq t_1$ such that for $t \geq t_2$, $z(t) > 0$, $w(t) > 0$, and $v(t) > 0$. For $t \geq t_2$, since

$$\begin{aligned}(a(t)v^{(n-1)}(t))' &= (a(t)w^{(n-1)}(t))' + p(a(t)w^{(n-1)}(t-\tau))' \\ &= (a(t)z^{(n-1)}(t))' + p(a(t)z^{(n-1)}(t-\tau))' \\ &\quad + p[(a(t)z^{(n-1)}(t-\tau))' + p(a(t)z^{(n-1)}(t-2\tau))'] \\ &= (a(t)z^{(n-1)}(t))' + 2p(a(t)z^{(n-1)}(t-\tau))' \\ &\quad + p^2(a(t)z^{(n-1)}(t-2\tau))' \\ &= -\delta q(t)f(x(t-\sigma)) - 2p\delta q(t)f(x(t-\tau-\sigma)) \\ &\quad - p^2\delta q(t)f(x(t-2\tau-\sigma))\end{aligned}$$

and

$$\begin{aligned}f(w(t-\sigma)) &= f(z(t-\sigma) + pz(t-\tau-\sigma)) \\ &\leq f(z(t-\sigma)) + pf(z(t-\tau-\sigma)) \\ &= f(x(t-\sigma) + px(t-\tau-\sigma)) + pf(x(t-\tau-\sigma) \\ &\quad + px(t-2\tau-\sigma)) \\ &\leq f(x(t-\sigma) + px(t-\tau-\sigma)) + pf(x(t-\tau-\sigma)) \\ &\quad + p^2 f(x(t-2\tau-\sigma))\end{aligned}$$

we obtain

$$\delta(a(t)v^{(n-1)})'(t) + q(t)f(w(t-\sigma)) \leq 0. \tag{44}$$

It follows, by applying Lemma 3.1 to the function $u(t)$ when $u(t) = z(t)$ or $u(t) = w(t)$ or $u(t) = v(t)$, that there are a $T > 0$ and an integer $l \in \{0, 1, ..., n-1\}$ with $(-1)^{n-l-1}\delta = 1$, such that for $t \geq T$

$$\begin{aligned}u^{(k)}(t) &\geq 0 \quad k = 0, 1, ..., l \\ (-1)^{n-k-1}u^{(k)}(t) &\geq 0 \quad k = l, ..., n-1.\end{aligned} \tag{45}$$

Clearly, $l = n$ is possible only if $\delta = -1$, and in this case we have $\lim_{t\to\infty} x(t) = \infty$. So it suffices to take $l \neq n$ whenever $\delta = -1$.

Now let $\delta = 1$ and n be even, or let $\delta = -1$ and n be odd. Then from (45), we see that $u(t)$ is increasing and $\delta u^{(n-1)}(t)$ is eventually positive. Therefore

$$f(w(s-\sigma)) \geq f(w(t_3 - \sigma)), \tag{46}$$

for $s \geq t_3$, since f is assumed to be increasing. Now we integrate (44) from t_3 to t to get

$$\delta a(t)v^{(n-1)}(t) - \delta a(t_3)v^{(n-1)}(t_3) + f(w(t_3 - \sigma)) \int_{t_3}^{t} q(s)\,ds \leq 0. \tag{47}$$

Because $\int^\infty q(t)\,dt = \infty$, we conclude from (47) that $\delta a(t)v^{(n-1)}(t) \to -\infty$ This is, of course, a contradiction with (45).

Let $\delta = 1$ and n be even, or let $\delta = -1$ and n be odd. If $l > 0$, the argument goes exactly the same as above and obtains a contradiction. If $l = 0$, then, integrating (44) from t_3 to ∞ and using (45) we get

$$-\delta a(t_3)v^{(n-1)}(t_3) + \int_{t_3}^{t} f(w(s-\sigma))q(s)\,ds \leq 0.$$

Thus,

$$\int_{t_3}^{\infty} f(w(s-\sigma))q(s)\,ds \le \delta a(t_3)v^{(n-1)}(t_3),$$

which implies that

$$\liminf_{t\to\infty} w(t) = 0.$$

Since $w(t)$ is monotone, this means that

$$\lim_{t\to\infty} w(t) = 0. \tag{48}$$

Using the fact that

$$0 \le x(t) \le z(t) \le w(t),$$

we conclude from (48) that

$$\lim_{t\to\infty} x(t) = 0.$$

This completes the proof of Theorem 3.4.

Remarks:

1. J. Graef, et al.[11] proved this theorem when $q(t)$ is τ periodic, $a(t) = 1$, $n = 2$, and $\delta = 1$. Thus the above theorem generalizes their results for arbitrary n in case of a more general equation (2).

2. Corollary 3.1 holds.

References

[1] D. D. Bainov and D. P. Mishev. *"Oscillation Theory for Neutral Differential Equations with Delay"*. IOP Publishing Ltd, Bristol, UK, 1992.

[2] R. Belman and K. L. Cooke. *"Differential-Difference Equations."* Academic Press, Inc., New York, 1963.

[3] R. K. Brayton and R. A. Willoughby. *On the numerical integration of a symmetric system of difference-differential equations of neutral type.* J.Math.Anal.Appl., 18 (1967), 182-189.

[4] Q. Chuanxi and G. Ladas. *Oscillation of neutral differential equations with variable coefficients.* Applicable Anal., 32 (1989), 215-228.

[5] R. S. Dahiya, T. Kusano, and M. Naito. *On almost oscillation of functional differential equations with deviating arguments.* J. Math. Anal. Appl., 98 (1984), 332-340.

[6] R. D. Driver. *Ordinary and Delay Differential Equations.* Springer-Verlag, New York, 1977.

[7] R. D. Driver. *Existence and continuous dependence of solutions of a neutral functional-differential equation.* Arch.Rational Mech.Anal., 19 (1965), 149-166.

[8] L. E. El'sgol'ts and S. B. Norkin. *"Introduction to the Theory and Application of Differential Equations with Deviating Arguments.* Academic Press, Inc., New York, 1973.

[9] K. E. Foster and R. C. Grimmer. *Nonoscillatory solutions of higher order delay equations.* J.Math.Anal.Appl., 77 (1980), 150-164.

[10] S. R. Grace and B. S. Lalli. *Oscillation of nonlinear second order neutral delay differential equations.* Rad. Mat., 3 (1987), 77-84.

[11] J. R. Graef, M. K. Grammatikopoulos, and P. W. Spikes. *Asymptotic properties of solutions of nonlinear neutral delay differential equations of the second order.* Rad. Mat., 4 (1988), 133-149.

[12] M. K. Grammatikopoulos and E. A. Grove. *Oscillation and asymptotic behavior of neutral differential equations with deviating arguments.* Applicable Anal., 22 (1986), 1-19.

[13] M. K. Grammatikopoulos, E. A. Grove, and G. Ladas. *Oscillation of first order neutral delay differential equations.* J.Math.Anal.Appl., 120 (1986), 510-520.

[14] M. K. Grammatikopoulos, G. Ladas, and A. Meimaridou. *Oscillation of second order neutral differential equations.* Rod.Mat., 2 (1985), 267-274.

[15] E. A. Grove and G. Ladas. *Oscillation of first order neutral differential equations.* J. Math. Anl. Appl., 120 (1986), 510-520.

[16] I. Gyori and G. Ladas. *"Oscillation Theory of Delay Differential Equations".* Oxford University Press, New York, 1991.

[17] J. Hale. *"Theory of Functional Differential Equations".* Springer-Verlag, New York, 1977.

[18] I. T. Kiguradze. *On the oscillation of solutions of equation* $d^m u/dt^m + a(t)u^m \mathrm{sgn} u = 0$. Mat.Sb., 65 (1964), 172-187.

[19] G. Ladas and Y. G. Sficas. *Oscillation of higher-order neutral equations.* J. Austral. Math. Soc., 27 (1986), 502-511.

[20] G. S. Ladde, V. Lakshmikantham and B. G. Zhang. *"Oscillation Theory of Differential Equations with Deviating Arguments".* Marcel Dekker, Inc., New York, 1987.

[21] A. S. Tantawy and R. S. Dahiya. *Oscillation of even order neutral delay differential equations. In "Differential equations stability and control".* Lecture notes in pure and applied mathematics. Marcel Dekker, Inc., 127 (1991), 503-509.

Third International Colloquium on Differential Equations pp. 219-227 (1993)
D. Bainov and V. Covachev (Eds)

Behaviour at infinity of solutions of systems of differential equations with delay.

J. DIBLÍK
Department of Math., Faculty of Electrical Engineering, Technical University of Brno, areál VUT Kraví Hora 21, 602 00 Brno, CZECHOSLOVAKIA, CZECH REPUBLIC.

Abstract – We introduce a method for establishment of existence of solutions of systems of differential equations with delay at infinity with asymptotic behaviour which is determined by means of solutions of some auxiliary systems of nondifferential equations. The method may be applied to various classes of retarded functional differential systems including some classes of nonlinear systems too.
Keywords – Retarded functional differential equation,topological method of T. Ważewski.

INTRODUCTION

In this paper we introduce a method for the establishment of existence of solutions of systems of differential equations with delay at infinity with asymptotic behaviour which is determined by means of solutions of some auxiliary systems of functional nondifferential equations. The method is shown on the case of systems of linear differential equations with delay

$$\dot{\boldsymbol{x}}(t) = -\boldsymbol{A}(t)\boldsymbol{x}(t-\tau) \tag{1}$$

where $\boldsymbol{x} \in \mathbb{R}^n, \boldsymbol{A}(t) = (a_{ij}(t))$ is $n \times n$ matrix on $I = [t_0, \infty), t_0 =$ const and $0 < \tau =$const.

Although the asymptotic properties of solutions of systems of linear differential equations with delay were investigated widely in connection with their numerous applications (e.g. by BAINOV and MISHEV [1], BELLMAN and COOKE [2], HALE [3], LAKSHMIKANTHAM and LEELA [4]) the considered case is not yet described.

The presented method may be applied to other classes of retarded functional differential systems including some classes of nonlinear systems too. Some related results are in the papers DIBLÍK [5, 6, 7] and ŠMARDA [8].

SOME PROPERTIES OF SOLUTIONS OF AUXILIARY SYSTEMS OF NONDIFFERENTIAL FUNCTIONAL EQUATIONS

Consider auxiliary system of linear nondifferential functional equations

$$\boldsymbol{\Delta}(t) = \boldsymbol{A}_\alpha(t)\boldsymbol{\Delta}(t-\tau) \tag{2}$$

where $\boldsymbol{\Delta} \in \mathbb{R}^n, \boldsymbol{A}_\alpha = \left(a^\alpha_{ij}(t)\right)$ is $n \times n$ continuous matrix on I such that $a^\alpha_{ij}(t) = \alpha_i(t)a_{ij}(t),\ i,j = 1,\dots,n$; $a_{ij}(t)$ are elements of matrix $\boldsymbol{A}(t)$ and $\alpha_i(t)$ are some given functions.

Definition 1. Let $\boldsymbol{A}_\alpha(t) \in C(I)$. The function $\boldsymbol{\Delta}(t) = \boldsymbol{\Delta}_1(t)$ is called *a solution of system* (2) if $\boldsymbol{\Delta}_1(t) \in C(I_1), I_1 = [t_0-\tau, \infty)$ and $\boldsymbol{\Delta}_1(t) \equiv \boldsymbol{A}_\alpha(t)\boldsymbol{\Delta}_1(t-\tau)$ on I.

Let $\boldsymbol{A}_\alpha(t) \in C^r(I)$. The function $\boldsymbol{\Delta}(t) = \boldsymbol{\Delta}_2(t)$ is called *a solution of system* (2) *of r-th class* if $\boldsymbol{\Delta}_2(t) \in C^r(I_1), r \geq 0$ and $\boldsymbol{\Delta}_2^{(r)}(t) \equiv [\boldsymbol{A}_\alpha(t)\boldsymbol{\Delta}_2(t-\tau)]^{(r)}$ on I.

Definition 2. The function $\boldsymbol{\nu}(t)$ is called *a initial function of r−th class* if $\boldsymbol{\nu}(t) \in C^r([t_0-\tau, t_0))$ and, moreover, the limit conditions

$$\boldsymbol{\nu}^{(s)}(t_0 - 0) \equiv [\boldsymbol{A}_\alpha(t_0)\boldsymbol{\nu}(t_0-\tau)]^{(s)} \tag{3}$$

hold for $s = 0, 1, \ldots, r$.

A solution of the system (2) of r−th class, $r \geq 0$, can be constructed by method of steps if $\boldsymbol{A}_\alpha(t) \in C^r(I)$ and if an initial function of r−th class $\boldsymbol{\nu}(t)$ is given. The coresponding solution of system (2) will be denoted as $\boldsymbol{\Delta}(t) \equiv \boldsymbol{\Delta}_\nu(t)$.

Lemma 1. *Let $\boldsymbol{\nu}(t)$ be an initial function of r−th class, $r \geq 0$ and $\boldsymbol{A}_\alpha \in C^r(I)$. Then there is a unique solution $\boldsymbol{\Delta}(t) = \boldsymbol{\Delta}_\nu(t)$ of r−th class of system* (2).

PROOF. Existence of such solution may be proved by method of steps. Indeed, we define

$$\boldsymbol{\Delta}_\nu^{(s)}(t) \equiv \boldsymbol{\Delta}_{\nu 0}^{(s)}(t) \equiv \boldsymbol{\nu}^{(s)}(t), \ s = 0, 1, \ldots, r; \ \ t \in [t_0-\tau, t_0)$$

and

$$\boldsymbol{\Delta}_\nu^{(s)}(t) \equiv \boldsymbol{\Delta}_{\nu k}^{(s)}(t), \ s = 0, 1, \ldots, r \ ; \ k = 1, 2, \ldots$$

on interval $t \in [t_0 + (k-1)\tau, t_0 + k\tau)$ where

$$\boldsymbol{\Delta}_{\nu k}^{(s)}(t) \equiv [\boldsymbol{A}_\alpha(t)\boldsymbol{\Delta}_{\nu k-1}(t-\tau)]^{(s)}.$$

A function constructed in such a way is the solution of system (2) of r−th class on I except, perhaps, of points $t_0 + (l-1)\tau$, $l = 1, 2, \ldots$. If $l = 1$ then, in view of (3), we obtain

$$\boldsymbol{\Delta}_\nu^{(s)}(t_0) = \boldsymbol{\Delta}_{\nu 1}^{(s)}(t_0) = [\boldsymbol{A}_\alpha(t_0)\boldsymbol{\Delta}_{\nu 0}(t_0-\tau)]^{(s)} =$$
$$= [\boldsymbol{A}_\alpha(t_0)\boldsymbol{\nu}(t_0-\tau)]^{(s)} = \boldsymbol{\nu}^{(s)}(t_0-0) = \boldsymbol{\Delta}_\nu^{(s)}(t_0-0), \ s = 0, 1, \ldots, r.$$

By method of induction we suppose that for $l = 1, 2, \ldots, m$, $m \geq 1$ the above mentioned property is valid. Then for $l = m+1$ we have

$$\boldsymbol{\Delta}_\nu^{(s)}(t_0+m\tau) = \boldsymbol{\Delta}_{\nu m+1}^{(s)}(t_0+m\tau) = [\boldsymbol{A}_\alpha(t_0+m\tau)\boldsymbol{\Delta}_{\nu 0}(t_0+(m-1)\tau)]^{(s)} =$$
$$= [\boldsymbol{A}_\alpha(t_0+m\tau)\boldsymbol{\Delta}_{\nu m-1}(t_0+(m-1)\tau-0)]^{(s)} = \boldsymbol{\Delta}_{\nu m}^{(s)}(t_0+m\tau-0) = \boldsymbol{\Delta}_\nu^{(s)}(t_0+m\tau-0)$$

and this property remains valid. The uniqueness of solution follows from the method of construction. The lemma is proved.

Let $\boldsymbol{\zeta} \in \mathbb{R}^n$ and $\boldsymbol{\eta} \in \mathbb{R}^n$ be two vectors. Then we define the relations of the type $\boldsymbol{\zeta} < \boldsymbol{\eta}$ etc. as relations between corresponding coordinates.

Consider two systems

$$\boldsymbol{\Xi}(t) = \boldsymbol{A}_\beta(t)\boldsymbol{\Xi}(t-\tau) \tag{4}$$

and

$$\boldsymbol{\Phi}(t) = \boldsymbol{A}_\gamma(t)\boldsymbol{\Phi}(t-\tau) \tag{5}$$

where $\boldsymbol{A}_\beta(t) = \left(a_{ij}^\beta(t)\right) = (\beta_i(t)a_{ij}(t))$, $\boldsymbol{A}_\gamma(t) = \left(a_{ij}^\gamma(t)\right) = (\gamma_i(t)a_{ij}(t))$, $i, j = 1, 2, \ldots, n$ and $\beta_i(t), \gamma_i(t)$ are some given functions positive on I.

Lemma 2. *Let $\boldsymbol{\xi}(t)$ and $\boldsymbol{\varphi}(t)$ be two initial functions of $r-$th class (if $\boldsymbol{A}_\alpha \equiv \boldsymbol{A}_\beta$ and $\boldsymbol{A}_\alpha \equiv \boldsymbol{A}_\gamma$ is put in (3) consequently) such that*

$$(6) \qquad 0 \le (-1)^s \boldsymbol{\xi}^{(s)}(t) < (-1)^{(s)} \boldsymbol{\varphi}^{(s)}(t), \ s = 0, 1, \dots, r; \ t \in [t_0 - \tau, t_0).$$

If, moreover, $\boldsymbol{A}_\beta(t) \in C^r(I)$, $\boldsymbol{A}_\gamma \in C^r(I)$, $(-1)^s \left(a_{ij}^\gamma(t)\right)^{(s)} \ge (-1)^s \left(a_{ij}^\beta(t)\right)^{(s)} \ge 0$, $0 < \beta_i(t) \le \gamma_i(t)$ and $\sum_{j=1}^{n} \left[(a_{ij}(t))^{(s)}\right]^2 > 0$ where $t \in I$, $s = 0, 1, \dots, r$; $i, j = 1, 2, \dots, n$ then for corresponding solutions of $r-$th class $\boldsymbol{\Xi}_\xi(t)$ of system (4) and $\boldsymbol{\Phi}_\varphi(t)$ of system (5) the following inequalities hold:

$$(7) \qquad 0 \le (-1)^s \boldsymbol{\Xi}_\xi^{(s)}(t) < (-1)^s \boldsymbol{\Phi}_\varphi^{(s)}(t), \ t \in I_1, \ s = 0, 1, \dots, r.$$

PROOF. The existence of solutions $\boldsymbol{\Xi}_\xi(t)$ and $\boldsymbol{\Phi}_\varphi(t)$ of $r-$th class follows from Lemma 1 if we put $\boldsymbol{\nu}(t) \equiv \boldsymbol{\xi}(t)$, $\boldsymbol{A}_\alpha(t) \equiv \boldsymbol{A}_\beta(t)$ and $\boldsymbol{\nu}(t) \equiv \boldsymbol{\varphi}(t)$, $\boldsymbol{A}_\alpha(t) \equiv \boldsymbol{A}_\gamma(t)$ in its formulation. The property (7) is valid for $s = 0$ on I because from (4), (5) on $[t_0, t_0 + \tau)$

$$\boldsymbol{\Phi}_\varphi(t) = \boldsymbol{A}_\gamma(t)\boldsymbol{\Phi}_\varphi(t - \tau) = \boldsymbol{A}_\gamma(t)\boldsymbol{\varphi}(t - \tau) > \boldsymbol{A}_\beta(t)\boldsymbol{\xi}(t - \tau) =$$
$$= \boldsymbol{A}_\beta(t)\boldsymbol{\Xi}_\xi(t - \tau) = \boldsymbol{\Xi}_\xi(t) \ge 0$$

and these inequalities may be re–proved on interval $[t_0 + k\tau, t_0 + (k+1)\tau)$ for each $k = 1, 2, \dots$. We will suppose according of the method of induction that (7) is valid for $s = 0, 1, \dots, l - 1 < r$. Then we obtain, on $[t_0, t_0 + \tau)$ by (6) and Leibniz formula,

$$(-1)^l \boldsymbol{\Phi}_\varphi^{(l)}(t) = (-1)^l \left[\boldsymbol{A}_\gamma(t)\boldsymbol{\Phi}_\varphi(t-\tau)\right]^{(l-k)} = (-1)^l \sum_{k=0}^{l} \binom{l}{k} \left[\boldsymbol{A}_\gamma(t)\right]^{(k)} \left[\boldsymbol{\Phi}_\varphi(t-\tau)\right]^{(l-k)} =$$

$$= \sum_{k=0}^{l} \binom{l}{k} \begin{bmatrix} \sum_{j=1}^{n} (-1)^k \left[a_{1j}^\gamma(t)\right]^{(k)} (-1)^{l-k} \left[\varphi_j(t-\tau)\right]^{(l-k)} \\ \cdots \\ \sum_{j=1}^{n} (-1)^k \left[a_{ij}^\gamma(t)\right]^{(k)} (-1)^{l-k} \left[\varphi_j(t-\tau)\right]^{(l-k)} \\ \cdots \\ \sum_{j=1}^{n} (-1)^k \left[a_{nj}^\gamma(t)\right]^{(k)} (-1)^{l-k} \left[\varphi_j(t-\tau)\right]^{(l-k)} \end{bmatrix} >$$

$$> l \sum_{k=0}^{l} \binom{l}{k} \begin{bmatrix} \sum_{j=1}^{n} (-1)^k \left[a_{1j}^\beta(t)\right]^{(k)} (-1)^{l-k} \left[\xi_j(t-\tau)\right]^{(l-k)} \\ \cdots \\ \sum_{j=1}^{n} (-1)^k \left[a_{ij}^\beta(t)\right]^{(k)} (-1)^{l-k} \left[\xi_j(t-\tau)\right]^{(l-k)} \\ \cdots \\ \sum_{j=1}^{n} (-1)^k \left[a_{nj}^\beta(t)\right]^{(k)} (-1)^{l-k} \left[\xi_j(t-\tau)\right]^{(l-k)} \end{bmatrix} =$$

$$= (-1)^l \sum_{k=0}^{l} \binom{l}{k} \left[\boldsymbol{A}_\beta(t)\right]^{(k)} \left[\boldsymbol{\Xi}_\xi(t-\tau)\right]^{(l-k)} = (-1)^l \left[\boldsymbol{A}_\beta(t)\boldsymbol{\Xi}_\xi(t-\tau)\right]^{(l)} = (-1)^l \boldsymbol{\Xi}_\xi^{(l)}(t) \ge 0$$

and these inequalities may be re–proved on interval $[t_0 + k\tau, t_0 + (k+1)\tau)$ for each $k = 1, 2, \dots$. The lemma is proved.

Lemma 3. *Let all assumptions of Lemma 2 be valid. Let there be functions $\boldsymbol{\psi}(t), \boldsymbol{\Psi}(t) \in C^r(I_1)$ such that for $s = 0, 1, \dots, r$*

$$(8_1) \qquad (-1)^s \boldsymbol{\psi}^{(s)}(t) < (-1)^s \left[\boldsymbol{A}_\beta(t)\boldsymbol{\psi}(t-\tau)\right]^{(s)}, \ t \in I,$$

$$(8_2) \qquad (-1)^s \left[\boldsymbol{A}_\gamma(t)\boldsymbol{\Psi}(t-\tau)\right]^{(s)} < (-1)^s \boldsymbol{\Psi}^{(s)}(t), \ t \in I$$

and, moreover,

$$(9_1) \qquad (-1)^s\boldsymbol{\psi}^{(s)}(t) < (-1)^s\boldsymbol{\xi}^{(s)}(t),\ t \in [t_0-\tau, t_0),$$

$$(9_2) \qquad (-1)^s\varphi^{(s)}(t) < (-1)^s\boldsymbol{\Psi}^{(s)}(t),\ t \in [t_0-\tau, t_0).$$

Then for the solution $\boldsymbol{\Xi}_\xi(t)$ *of* (4) *and* $\boldsymbol{\Phi}_\varphi$ *of* (5)

$$(10) \qquad (-1)^s\boldsymbol{\psi}^{(s)}(t) < (-1)^s\boldsymbol{\Xi}_\xi^{(s)}(t) < (-1)^s\boldsymbol{\Phi}_\varphi^{(s)}(t) < (-1)^s\boldsymbol{\Psi}^{(s)}(t),$$

hold on I_1.

PROOF. Let (8_1) and (9_1) hold, $s=0$ and $t \in [t_0, t_0+\tau)$. Then

$$\boldsymbol{\Xi}_\xi(t) = \boldsymbol{A}_\beta(t)\boldsymbol{\Xi}_\xi(t-\tau) = \boldsymbol{A}_\beta(t)\boldsymbol{\xi}(t-\tau) > \boldsymbol{A}_\beta(t)\boldsymbol{\psi}(t-\tau) > \boldsymbol{\psi}(t)$$

and these inequalities may be re–proved by analogy on interval $[t_0+k\tau, t_0+(k+1)\tau)$ for each $k=1,2,\ldots$. Let (8_1) be valid for $s=0,1,\ldots,l-1<r$. Then on $[t_0, t_0+\tau)$

$$(-1)^l\boldsymbol{\Xi}_\xi^{(l)}(t) = (-1)^l\left[\boldsymbol{A}_\beta(t)\boldsymbol{\Xi}_\xi(t-\tau)\right]^{(l)} = (-1)^l\sum_{k=0}^{l}\binom{l}{k}\left[\boldsymbol{A}_\beta(t)\right]^{(k)}\boldsymbol{\Xi}_\xi^{(l-k)}(t-\tau) =$$

$$= \sum_{k=0}^{l}\binom{l}{k}\begin{bmatrix} \sum_{j=1}^{n}(-1)^k\left[a_{1j}^{\beta}(t)\right]^{(k)}(-1)^{l-k}\xi_j^{(l-k)}(t-\tau) \\ \cdots \\ \sum_{j=1}^{n}(-1)^k\left[a_{ij}^{\beta}(t)\right]^{(k)}(-1)^{l-k}\xi_j^{(l-k)}(t-\tau) \\ \cdots \\ \sum_{j=1}^{n}(-1)^k\left[a_{nj}^{\beta}(t)\right]^{(k)}(-1)^{l-k}\xi_j^{(l-k)}(t-\tau) \end{bmatrix} >$$

$$> \sum_{k=0}^{l}\binom{l}{k}\begin{bmatrix} \sum_{j=1}^{n}(-1)^k\left[a_{1j}^{\beta}(t)\right]^{(k)}(-1)^{l-k}\psi_j^{(l-k)}(t-\tau) \\ \cdots \\ \sum_{j=1}^{n}(-1)^k\left[a_{ij}^{\beta}(t)\right]^{(k)}(-1)^{l-k}\psi_j^{(l-k)}(t-\tau) \\ \cdots \\ \sum_{j=1}^{n}(-1)^k\left[a_{nj}^{\beta}(t)\right]^{(k)}(-1)^{l-k}\psi_j^{(l-k)}(t-\tau) \end{bmatrix} =$$

$$= (-1)^l\left[\boldsymbol{A}_\beta(t)\boldsymbol{\psi}(t-\tau)\right]^{(l)} > (-1)^l\boldsymbol{\psi}^{(l)}(t)$$

and these inequalities may be re–proved on interval $[t_0+k\tau, t_0+(k+1)\tau)$ for each $k=1,2,\ldots$. Therefore the left sides of inequalities (10) hold. By analogy we can prove that the right sides of (10) hold in the case if (8_2) and (9_2) hold. The inequalities between $\boldsymbol{\Xi}_\xi(t)$ and $\boldsymbol{\Phi}_\varphi(t)$ follow from Lemma 2.

MAIN RESULTS

Consider the system of functional differential equations of the retarded type

$$(11) \qquad \dot{\mathbf{y}}(t) = \mathbf{f}(t, \mathbf{y}_t)$$

where $\mathbf{y} \in \mathbb{R}^n$, $\mathbf{y}_t$ is an element from the space of continuous function $K = K([-\tau, 0], \mathbb{R}^n)$, $\mathbf{y}_t(\theta) = \mathbf{y}(t+\theta)$ where $\theta \in [-\tau, 0]$, $\mathbf{f} : \Omega \to \mathbb{R}^n$, $\Omega = (t_0 - \tau - \varepsilon, +\infty) \times K$, $0 < \varepsilon$,

ε =const and $\mathbf{f}$ is a continuous mapping such that each element $(\delta, \boldsymbol{\pi}) \in \Omega$ determines a unique solution $\mathbf{y}(\delta, \boldsymbol{\pi})$ on its maximal interval $D_{\delta,\boldsymbol{\pi}} = [\delta, a), \delta < a \leq \infty$. The elements $(t, \boldsymbol{\pi}) \in \mathbb{R} \times K$ are assumed to be such that $(t, \boldsymbol{\pi}) \in \Omega$.

Let $\omega_0 = \{(t, \mathbf{y}) \in \mathbb{R} \times \mathbb{R}^n, t \in I_1\}$ and

$$\omega_1 = \{(t, \mathbf{y}) \in \omega_0, l_i(t, \mathbf{y}) < 0, i = 1, 2, \dots, p,\ m_j(t, \mathbf{y}) < 0, j = 1, 2, \dots, q, p^2 + q^2 > 0\}$$

where the functions $l_i(t, \mathbf{y}),\ i = 1, 2, \dots, p,\ \ m_j(t, \mathbf{y}), j = 1, 2, \dots, q$ are defined as

$$l_i(t, \mathbf{y}) \equiv (y_i - \varrho_i(t))(y_i - \delta_i(t)),$$
$$m_j(t, \mathbf{y}) \equiv (y_{p+j} - \varrho_{p+j}(t))(y_{p+j} - \delta_{p+j}(t)),$$

functions $\varrho_k(t), \delta_k(t),\ k = 1, 2, \dots, p+q$ are continuous on I_1, countinuously differentiable on I and $\varrho_k(t) < \delta_k(t)$ on I_1.

The following theorem, which we will use in our investigations, can be proved by analogy with the proof of Theorem 2 from DIBLÍK [6] which is based on the topological method of T. Ważewski in the form of RYBAKOWSKI [9]. Therefore its proof is omitted.

Theorem 1. *Let*

a) for each $i \in \{1, 2, \dots, p\}$; $(t, \mathbf{y}) \in \partial\omega_1$ *if* $l_i(t, \mathbf{y}) = 0$ *and* $\boldsymbol{\pi} \in K$, $\boldsymbol{\pi}(0) = \mathbf{y}$, $(t + \theta, \boldsymbol{\pi}(\theta)) \in \omega_1$, $\theta \in [-\tau, 0)$ *hold on* I:

$$\delta_i' < f_i(t, \boldsymbol{\pi}) \text{ if } y_i = \delta_i(t) \tag{12}$$

and

$$\varrho_i'(t) > f_i(t, \boldsymbol{\pi}) \text{ if } y_i = \varrho_i(t); \tag{13}$$

b) for each $j \in \{1, 2, \dots, q\}$; $(t, \mathbf{y}) \in \partial\omega_1$ *if* $m_j(t, \mathbf{y}) = 0$ *and* $\boldsymbol{\pi} \in K$, $\boldsymbol{\pi}(0) = \mathbf{y}$, $(t + \theta, \boldsymbol{\pi}(\theta)) \in \omega_1$, $\theta \in [-\tau, 0)$ *hold on* I:

$$\varrho_{p+j}'(t) < f_{p+j}(t, \boldsymbol{\pi}) \quad \text{if} \quad y_{p+j} = \varrho_{p+j}(t)$$

and

$$\delta_{p+j}' > f_{p+j}(t, \boldsymbol{\pi}) \quad \text{if} \quad y_{p+j} = \delta_{p+j}(t).$$

Then there is a noncountable set of solutions of (11) $\mathbf{y} = \mathbf{y}(t)$ *such that on* I_1 *(including the values of corresponding initial functions)*

$$\varrho_i(t) < y_i(t) < \delta_i(t),\ i = 1, 2, \dots, n. \tag{14}$$

Theorem 2. *Let all conditions of Lemma 2 hold for* $r = 1$. *Let, moreover, be for* $t \in I$, $i = 1, 2, \dots, n$

$$\gamma_i(t)\Phi_{\varphi_i}'(t) + \Phi_{\varphi_i}(t) \leq 0, \tag{15}$$

$$\beta_i(t)\Xi_{\xi_i}'(t) + \Xi_{\xi_i}(t) \geq 0 \tag{16}$$

where $\boldsymbol{\Xi}_{\boldsymbol{\xi}}(t)$ *and* $\boldsymbol{\Phi}_{\boldsymbol{\varphi}}(t)$ *are solutions of systems* (4) *and* (5).
Then there is a noncountable set of solutions $\mathbf{x} = \mathbf{x}(t)$ *of system* (1) *such that on* I_1

$$0 < (-1)^s \boldsymbol{\Xi}_{\boldsymbol{\xi}}^{(s)}(t) < (-1)^s \mathbf{x}^{(s)}(t) < (-1)^s \boldsymbol{\Phi}_{\boldsymbol{\varphi}}^{(s)}(t), \tag{17}$$

$s = 0, 1$.

PROOF. We will apply Theorem 1 and Lemma 2. Let $p = n$, $q = 0$, $\varrho_i(t) \equiv \Xi_{\xi_i}(t)$, $\delta_i(t) \equiv \Phi_{\varphi_i}(t)$, $y_i \equiv x_i$, $i = 1,2,\ldots,n$ and $\mathbf{f}(t,\mathbf{x}_t) \equiv -\mathbf{A}(t)\mathbf{x}(t-\tau)$. We verify the inequalities (12). In view of (15) we have

$$\delta_i'(t) - f_i(t,\boldsymbol{\pi}) = \Phi_{\varphi_i}'(t) + \sum_{j=1}^{n} a_{ij}(t)\pi_j(t-\tau) <$$

$$< \Phi_{\varphi_i}'(t) + \sum_{j=1}^{n} a_{ij}(t)\Phi_{\varphi_i}(t-\tau) = \Phi_{\varphi_i}'(t) + \frac{1}{\gamma_i(t)}\sum_{j=1}^{n}\gamma_i(t)a_{ij}(t)\Phi_{\varphi_i}(t-\tau) =$$

$$= \Phi_{\varphi_i}'(t) + \frac{1}{\gamma_i(t)}\Phi_{\varphi_i}(t) \leq 0$$

and $\delta_i'(t) - f_i(t,\boldsymbol{\pi}) < 0$ on I.

By analogy, in view of (16), we can verify the inequalities (13). Therefore, as it follows from (7) and (14), the inequalities (17) for $s = 0$ hold. At the end we prove the inequalities (17) for $s = 1$. For $t \in I$ and $i = 1,2,\ldots,n$ we have, in view of (15),

$$x_i'(t) - \Phi_{\varphi_i}'(t) = -\sum_{j=1}^{n} a_{ij}(t)x_j(t-\tau) - \Phi_{\varphi_i}'(t) >$$

$$> -\sum_{j=1}^{n} a_{ij}(t)\Phi_{\varphi_j}(t-\tau) - \Phi_{\varphi_i}'(t) = \frac{-\Phi_{\varphi_i}(t)}{\gamma_i(t)} - \Phi_{\varphi_i}'(t) \geq 0.$$

By (16) we can perform analogically that $x_i'(t) - \Xi_{\xi_i}'(t) < 0$. The application of (7) finishes the proof.

Theorem 3. *Let* $\mathbf{A}(t) \in C^1(I)$, $(-1)^s(a_{ij}(t))^{(s)} \geq 0$, $\sum_{j=1}^{n}\left[(a_{ij}(t))^{(s)}\right]^2 > 0$ *and* $\int^{\infty} a_{ij}(t)dt < \infty, s = 0,1$; $t \in I$, $i,j = 1,2,\ldots,n$. *Let, moreover,* $\boldsymbol{\varphi}(t)$ *be an initial function of 1-th class,* $0 < (-1)^s\boldsymbol{\varphi}^{(s)}(t)$, $s = 0,1$ *and* $a_{ij}(t) + (\alpha_i(t)a_{ij}(t))' \leq 0$ *where*

$$\alpha_i(t) \equiv \frac{K}{\sum_{j=1}^{n} a_{ij}(t)}\int_t^{\infty}\left[\sum_{j=1}^{n} a_{ij}(s)\right] ds,\ 1 < K = \text{const},\ t \in I,\ i,j = 1,2,\ldots,n.$$

Then there is a noncountable set of solutions $\mathbf{x} = \mathbf{x}(t)$ *of* (1) *such that*

$$0 < (-1)^{(s)}\mathbf{x}^{(s)}(t) < (-1)^{(s)}\Phi_{\varphi}^{(s)}(t),\ s = 0,1$$

on I_1.

PROOF. It is not difficult to verify that the conditions of Lemma 2 are valid if we put $\boldsymbol{\xi}(t) \equiv 0$, $\alpha = \beta = \gamma$, $\alpha_i(t) \equiv \beta_i(t) \equiv \gamma_i(t)$ and $r = 1$ in its formulation. Obviously in such case the solution $\Xi_{\boldsymbol{\xi}}(t)$ of system (4) is trivial, i.e. $\Xi_{\boldsymbol{\xi}}(t) \equiv 0$. Now we verify the conditions of Theorem 2. Inequalities (16) are valid. Further on I

$$\gamma_i(t)\Phi_{\varphi_i}'(t) + \Phi_{\varphi_i}(t) =$$

$$= \gamma_i(t)\left[\sum_{j=1}^{n}\left[\gamma_i(t)a_{ij}(t)\Phi_{\varphi_i}(t-\tau)\right]'\right] + \sum_{j=1}^{n}\gamma_i(t)a_{ij}(t)\Phi_{\varphi_i}(t-\tau) =$$

$$= \sum_{j=1}^{n}\gamma_i(t)\left[\left[a_{ij}(t) + (\gamma_i(t)a_{ij}(t))'\right]\Phi_{\varphi_i}(t-\tau) + \gamma_i(t)a_{ij}(t)\Phi_{\varphi_i}'(t-\tau)\right] \leq 0,$$

$i = 1,2,\ldots,n$ and inequalities (15) are valid too. Then the conclusion of Theorem 3 follows from the conclusion of Theorem 2.

Theorem 4. *Let there be function* $\boldsymbol{\psi}(t), \boldsymbol{\Psi}(t) \in C^1(I_1)$ *for which the inequalities* (8_1), (8_2) *hold if* $s = 0$; $0 < \beta_i(t) \leq \gamma_i(t)$, $0 \leq a_{ij}(t)$, $\sum_{j=1}^{n} a_{ij}^2(t) > 0$, $i, j = 1, 2, \ldots, n$, $t \in I$. *Let, moreover,*

$$\gamma_i(t)\Psi_i'(t) + \Psi_i(t) \leq 0, \tag{18}$$

$$\beta_i(t)\psi_i'(t) + \psi_i(t) \geq 0 \tag{19}$$

and $(-1)^s\boldsymbol{\psi}^{(s)}(t) < (-1)^s\boldsymbol{\Psi}^{(s)}(t)$, $s = 0, 1$ *for* $t \in I$, $i = 1, 2, \ldots, n$.
Then there is a noncountable set of solutions $\mathbf{x} = \mathbf{x}(t)$ *of* (1) *such that*

$$(-1)^s\boldsymbol{\psi}^{(s)}(t) < (-1)^s\mathbf{x}^{(s)}(t) < (-1)^s\boldsymbol{\Psi}^{(s)}(t), \quad s = 0, 1 \tag{20}$$

on I_1.

PROOF. The proof can be made by analogy with the proof of Theorem 2. If we put $p = n$, $q = 0$, $\varrho_i(t) \equiv \psi_i(t)$, $\delta_i(t) \equiv \Psi_i(t)$, $y_i \equiv x_i$, $i = 1, 2, \ldots, n$, $\mathbf{f}(t, \mathbf{x}_t) \equiv -\mathbf{A}(t)\mathbf{x}(t-\tau)$ and $t \in I$ then

$$\delta_i'(t) - f_i(t, \boldsymbol{\pi}) = \Psi_i'(t) + \sum_{j=1}^{n} a_{ij}(t)\pi_j(t-\tau) <$$
$$< \Psi_i'(t) + \sum_{j=1}^{n} a_{ij}(t)\Psi_j(t-\tau) < \Psi_i'(t) + \frac{1}{\gamma_i(t)}\Psi_i(t) \leq 0$$

and by analogy we can show that $\varrho_i'(t) - f_i(t, \boldsymbol{\pi}) > 0$. Further we obtain

$$x_i'(t) - \Psi_i'(t) = -\sum_{j=1}^{n} a_{ij}(t)x_j(t-\tau) - \Psi_i'(t) >$$
$$> -\sum_{j=1}^{n} a_{ij}(t)\Psi_j(t-\tau) - \Psi_i'(t) > -\Psi_i'(t) + \frac{-1}{\gamma_i(t)}\Psi_i(t) \geq 0$$

for $i = 1, 2, \ldots, n$, $t \in I$ and by analogy we can show that $x_i'(t) - \psi_i'(t) < 0$. The theorem is proved.

Remark 1. From Lemma 3 and Theorem 4 it follows that the solutions $\boldsymbol{\Xi}_\xi(t)$, $\boldsymbol{\Phi}_\varphi(t)$ and $\mathbf{x}(t)$ satisfy the same type inequalities on I.

Remark 2. Often there may be put

$$\psi_i(t) \equiv N_i \exp\left[-\int_{t_0}^{t} \frac{ds}{\beta_i(s)}\right], \Psi_i(t) \equiv M_i \exp\left[-\int_{t_0}^{t} \frac{ds}{\gamma_i(s)}\right],$$

$i = 1, 2, \ldots, n$ where $0 < N_i < M_i$; $N_i, M_i =$const. Then the inequalities (18), (19) hold. The functions $\beta_i(t), \gamma_i(t), i = 1, 2, \ldots, n$ may be frequently taken in the form

$$\beta_i(t) \equiv -\tilde{N}_i\left[\ln(K\sum_{j=1}^{n} a_{ij}(t))\right]^{-1}, \gamma_i(t) \equiv -\tilde{M}_i\left[\ln(K\sum_{j=1}^{n} a_{ij}(t))\right]^{-1}$$

where $\tilde{N}_i$, $\tilde{M}_i$ and K are positive constants and $\tilde{N}_i < \tilde{M}_i$.

EXAMPLES

Example 1. Consider the scalar equation

$$\dot{x}(t) = -\left[\exp(-t)\right] x(t-1).$$

If we put (in correspondence with Remark 2) $N = 0.5, M = 1, K = 1, \beta = 0.5t^{-1}$, $\gamma(t) = 2t^{-1}$,$t_0 = 10, \tau = 1, \psi(t) = 0.5\exp(-t^2 + 100), \Psi(t) = \exp\left[0.25(-t^2 + 100)\right]$ then all assumptions of Theorem 4 are valid. Therefore there is a noncountable set of solutions $x = x(t)$ such that the inequalities (20) hold.

Example 2. Consider the system

$$\dot{x}_1(t) = -e^{-5t}x_1(t-1) - e^{-2t}x_2(t-1),$$
$$\dot{x}_2(t) = -e^{-4t}x_1(t-1) - e^{-2t}x_2(t-1).$$

If we put

$$\beta_1(t) = \left[\exp(-3t^2 + 3t - 1)\right]\left[\exp(-5t) + \exp(-2t)\right]^{-1},$$
$$\beta_2(t) = \left[\exp(-3t^2 + 3t - 1)\right]\left[\exp(-4t) + \exp(-2t)\right]^{-1},$$
$$\gamma_1(t) = \left[\exp(-2t + 1)\right]\left[\exp(-5t) + \exp(-2t)\right]^{-1},$$
$$\gamma_2(t) = \left[\exp(-2t + 1)\right]\left[\exp(-4t) + \exp(-2t)\right]^{-1}$$

and $\tau = 1$ then the systems (4), (5) have the solutions: $\Xi_{\xi}(t) = \left(e^{-t^3}, e^{-t^3}\right)^T$, $\Phi_{\varphi}(t) = \left(e^{-t^2}, e^{-t^2}\right)^T$ correspondingly where $\xi(t) = \left(e^{-t^3}, e^{-t^3}\right)^T$ and $\varphi(t) = \left(e^{-t^2}, e^{-t^2}\right)^T$. Then all assumptions of Theorem 2 are valid for sufficiently large $t_0 \in \mathbb{R}$ and there is a noncountable set of solutions $\mathbf{x} = \mathbf{x}(t)$ such that the inequalities (17) hold.

References

1. D.D. Bainov and D.P. Mishev. *Oscillation Theory for Neutral Differential Equations with Delay.* Adam Hilger, Bristol, Philadelphia and New York (1991).

2. R. Bellman and K.L. Cooke. *Differential-difference Equations.* Academic Press, New York and London (1963).

3. J. Hale. *Theory of Functional Differential Equations.* Springer-Verlag, New York, Heidelberg, Berlin (1977).

4. V. Lakshmikantham and S. Leela. *Differential and Integral Inequalities.* Vol. 2. Academic Press, (1969).

5. J. Diblík. Asymptotic behaviour of solutions of linear differential equations with delay, *Ann. Polon. Math.*, in press.

6. J. Diblík. On asymptotic behaviour of solutions of some classes of retarded functional diffrential equations, *Izv. Vyssh. Uchebn. Zaved., Mat.*,15-20 (1991) (Russian).

7. J. Diblík. Existence of solutions with asymptotic behaviour of a certain type of some systems of functional differential equations with delay, *Siberian. Math. J.*, 32, No2, 222-226(1991); *Sibirsk. Mat. Zh.*, 32, 55-60 (1991).

8. Z. Šmarda. The asymptotic behaviour of solutions of the singular integro–differential equation, *Demonstratio Mathematica*, XXII, 293–308 (1989).

9. K. P. Rybakowski. Ważewski's principle for retarded functional differential equations, *J. Differential Equations*, 36, 117–138 (1980).

Bei Fragen zur Produktsicherheit wenden Sie sich bitte an:
If you have any questions regarding product safety,
please contact:

Walter de Gruyter GmbH
Genthiner Straße 13
10785 Berlin
productsafety@degruyterbrill.com